THE ANSWER KEY

A Comprehensive Explanation of Problem Solving
Methods for General Chemistry Success

Volume 2

First Edition

Rachel Turoscy

Middlesex County College

Bassim Hamadeh, CEO and Publisher
John Remington, Senior Field Acquisitions Editor
Gem Rabanera, Project Editor
Chelsey Schmid, Production Editor
Emely Villavicencio, Senior Graphic Designer
Trey Soto, Licensing Associate
Natalie Piccotti, Senior Marketing Manager
Kassie Graves, Vice President of Editorial
Jamie Giganti, Director of Academic Publishing

Copyright © 2019 by Cognella, Inc. All rights reserved. No part of this publication may be reprinted, reproduced, transmitted, or utilized in any form or by any electronic, mechanical, or other means, now known or hereafter invented, including photocopying, microfilming, and recording, or in any information retrieval system without the written permission of Cognella, Inc. For inquiries regarding permissions, translations, foreign rights, audio rights, and any other forms of reproduction, please contact the Cognella Licensing Department at rights@cognella.com.

Trademark Notice: Product or corporate names may be trademarks or registered trademarks, and are used only for identification and explanation without intent to infringe.

Cover image copyright© 2016 iStockphoto LP/Cakeio.

Printed in the United States of America.

ISBN: 978-1-5165-3028-1 (pbk) / 978-1-5165-3029-8 (br)

Table of Contents

To the Student v

Chapter 1. Liquids, Solids, and Phase Changes 1
   Section 1.1 Solids: Crystalline Density Calculations 2
   Section 1.2 Vapor Pressure: Clausius-Clapeyron Equation 5
      Section 1.2.1 Plotting and Linear Relationships 5
      Section 1.2.2 Pressure and Temperature – Clausius-Clapeyron Equation 5
   Section 1.3 Phase Transitions: Energy Calculations 9
   Credits 14

Chapter 2. Solutions 15
   Section 2.1 Henry's Law 16
   Section 2.2 Concentration Expressions and Calculations 17
      Section 2.2.1 Molarity (M) 17
      Section 2.2.2 Molality 18
      Section 2.2.3 Percent by Mass, ppm, ppb 18
      Section 2.2.4 Mole Fraction 20
      Section 2.2.5 Converting between Concentration Units 21
   Section 2.3 Colligative Properties 24
      Section 2.3.1 Vapor Pressure Lowering 24
         Section 2.3.1.1 Non-Volatile Solute 24
         Section 2.3.1.2 Volatile Solute 25
      Section 2.3.2 Boiling Point Elevation and Freezing Point Lowering 27
      Section 2.3.3 Osmotic Pressure 29
      Section 2.3.4 Colligative Properties and Ionic Solutions, van 't Hoff Factor 30
         Section 2.3.4.1 van 't Hoff Factor Determination 31
      Section 2.3.5 Additional Colligative Property Problems, Calculation of Molar Mass 32

Chapter 3. Kinetics 35
   Section 3.1 Determining the Rate of a Reaction via Stoichiometry 36

Section 3.2 Dependence of Rate on Concentration; Rate Law Determination 38
Section 3.3 Integrated Rate Laws 43
    Section 3.3.1 First Order Problem 44
    Section 3.3.2 Second Order Problem 45
Section 3.4 Temperature and the Effect on Rate: The Arrhenius Equations 47
Section 3.5 Reaction Mechanisms 51

Chapter 4. Equilibria 55
Section 4.1 Determination of K_c and K_p, Equilibrium Concentrations 56
Section 4.2 The Relationship between K_c and K_p 59
Section 4.3 Equilibrium Constant Manipulation and the Chemical Equation 61
Section 4.4 The Use of the Reaction Quotient, Q 64
Section 4.5 Finding Equilibrium Concentrations, ICE Tables 66
Section 4.6 Le Chatelier's Principle 87

Chapter 5. Acids and Bases 91
Section 5.1 K_w Equilibrium, pH and pOH Problems 92
Section 5.2 Strong Acids and Strong Bases 95
Section 5.3 Weak Acids and Bases 97
Section 5.4 pH of Salts in Solution 108
    Section 5.4.1 K_a, K_b, and K_w 108
    Section 5.4.2 pH of Salt Solutions 109
        Section 5.4.2.1 Salts from Strong/Strong Acid/Base Systems 109
        Section 5.4.2.2 Salts from Weak and Strong Acid/Base Systems 110
        Section 5.4.2.3 Salts from Weak and Weak Acid/Base Systems 111
        Section 5.4.2.4 pH Determination of a Salt Solution 111

Chapter 6. Buffers 115
Section 6.1 The Common Ion Effect 116
Section 6.2 Buffers 120
    Section 6.2.1 Common Ion Effect and Henderson-Hasselbalch Equations 120
Section 6.3 Calculations for the Addition of Acids or Bases to Buffers 125
    Section 6.3.1 Addition of a Strong Acid 125

Section 6.3.2 Addition of a Strong Base 128

Section 6.4 Forming a Buffer with a Specific pH 131

Chapter 7. Titrations 133

Section 7.1 Strong Acid/Strong Base Titrations... 135

Section 7.1.1 Initial State – No Titrant – Strong Acid/Strong Base 135

Section 7.1.2 Before the Equivalence Point – Strong Acid/Strong Base 136

Section 7.1.3 At the Equivalence Point – Strong Acid/Strong Base 138

Section 7.1.4 After the Equivalence Point – Strong Acid/Strong Base................ 140

Section 7.2 Weak Acid/Strong Base Titrations 143

Section 7.2.1 Initial State – No Titrant – Weak Acid/Strong Base....................... 143

Section 7.2.2 Before the Equivalence Point – Weak Acid/Strong Base 146

Section 7.2.3 At the Equivalence Point – Weak Acid/Strong Base..................... 148

Section 7.2.4 After the Equivalence Point – Weak Acid/Strong Base 152

Section 7.3 Weak Base/Strong Acid Titrations 155

Section 7.3.1 Initial State – No Titrant – Weak Base/Strong Acid....................... 155

Section 7.3.2 Before the Equivalence Point – Weak Base/Strong Acid 158

Section 7.3.3 At the Equivalence Point – Weak Base/Strong Acid..................... 160

Section 7.3.4 After the Equivalence Point – Weak Base/Strong Acid 165

Section 7.4 Typical Problems and Student Problems.. 168

Chapter 8. Solubility Equilibria 173

Section 8.1 Determination of K_{sp} 174

Section 8.2 Precipitation Problems 181

Chapter 9. Thermodynamics II.. 185

Section 9.1 Entropy and the Second Law of Thermodynamics 186

Section 9.1.1 Entropy of a Reaction, $\Delta S^o_{rxn} = \Sigma S^o_{f\,products} - \Sigma S^o_{f\,reactants}$ 187

Section 9.1.2 Predicting Spontaneity with $\Delta S_{universe} = \Delta S_{system} + \Delta S_{surroundings}$ 188

Section 9.1.3 Predicting Entropy for a Phase Change 189

Section 9.2 Gibbs Free Energy, G 191

Section 9.2.1 Gibbs Free Energy of a Reaction, $\Delta G^o_{rxn} = \Sigma G^o_{f\,products} - \Sigma G^o_{f\,reactants}$ 191

Section 9.2.2 Gibbs Free Energy of a Reaction; $\Delta G° = \Delta H° - T\Delta S°$ 192

Section 9.2.2.1 Type 1 Problem: Calculation of ΔG° 192

Section 9.2.2.2 Type 2 Problem: Calculation of ΔG° at Different Temperatures............ 194

Section 9.2.2.3 Type 3 Problem: Calculation of Temperature for Reaction Spontaneity 195

Section 9.2.3 Gibbs Free Energy and Equilibrium Constants 196

Chapter 10. Electrochemistry 199

Section 10.1 Galvanic Cells: Cell Potential, E^o_{cell} or Electromotive Force, EMF 201

Section 10.2 Gibbs Free Energy and Cell Potential 205

Section 10.3 Nernst Equation, Non-Standard Conditions 209

Section 10.4 Electrolysis Calculations 211

To the Student

Mastering chemistry has always proven an arduous task for many students pursuing studies in any natural science or engineering. After teaching in a community college for over a decade, I realized that many students need a more comprehensive explanation for solving chemistry problems encountered in the first two semester of any general chemistry course. In every section of this book, I have completely outlined my approach to solving problems and the mathematics that are necessary to solve the problem. I have attempted to simplify concepts and then provide a strategy to solving problems. As you will see, I used the classic linear approach to solutions and conversions, and I have also introduced a new stacking method that eliminates the use of fractions. Each example I give is very detailed, and I am hopeful that using my approach and strategies will help all students as they progress through these first two semesters.

Chapter 1. Liquids, Solids, and Phase Changes

Most textbooks begin this section with a discussion of intermolecular forces and their relevance to the liquid and solid phases. The three main areas in which mathematical calculations and problem solving are relevant in this chapter involve vapor pressure, density, and energy calculations for phase transitions.

Section 1.1 Solids: Crystalline Density Calculations

There are many different types of unit cells that make up the crystal lattices for compounds or elements. This chapter presents an explanation of simple cubic structures. The cubic system has three possible cells: Simple Primitive Cubic, Body Center Cubic, and Face Center Cubic.

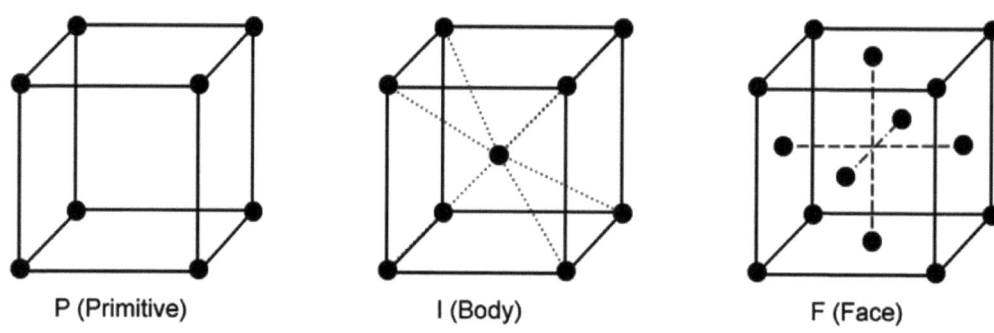

Figure 1.1

http://biochem.co/2008/08/crystal-structure-studies/

A primitive cell consists of 1 atom, the body center consists of 2 atoms, and the face center cubic structure consists of 4 atoms. You will need to remember these values are to solve density calculations with this equation:

$$\text{DENSITY} = \frac{\frac{(\text{number of atoms in the unit cell})(\text{Molar Mass})}{\text{Avogadro's Number}}}{(\text{edge length})^3}$$

Three additional equations relate edge length (a) with radius (r) in the three systems:

Simple Cubic: a = 2r   Body Centered Cubic: a = 4r/√3   Face Centered Cubic: a = 2√2r

Sample Problem 1.1 Crystal Density Calculations

Molybdenum crystallizes with a body centered unit cell. The edge length of the cell is 314 pm. What is the density of molybdenum?

Solution:

Step 1. Select the correct equation for solution. Since the question asks for density, we use this equation:

$$\text{DENSITY} = \frac{\frac{(\text{number of atoms in the unit cell})(\text{Molar Mass})}{\text{Avogadro's Number}}}{(\text{edge length})^3}$$

Step 2. What do the questions ask for, and what variables do we know?

Underline the unknown variable (what is asked for) and then circle the known quantities:

Molybdenum crystallizes with a ⭕body centered unit cell⭕. The ⭕edge length of the cell is 314 pm.⭕ What is the density of molybdenum?

Step 3a. Convert pm to cm for edge length. Density for solids has the units of grams/cm³

$$\frac{314\ pm}{1} \times \frac{1\ cm}{1 \times 10^{10}\ pm} = 314 \times 10^{-10}\ cm$$

Step 3b. Fill in the quantities in equation. The molar mass of Mo is 95.96 grams/mole, and a body centered cubic system has 2 atoms per cell. Avogadro's # = 6.02 x 10²³ atoms/mole.

$$\text{DENSITY} = \frac{\frac{(2\ atoms)(95.96\ grams/mole)}{6.02 \times 10^{23}\ atoms/mole}}{(314 \times 10^{-10}\ cm)^3}$$

Unit Cancellation:

$$\text{DENSITY} = \frac{\frac{(2\ \cancel{atoms})(95.96\ grams/\cancel{mole})}{6.02 \times 10^{23}\ \cancel{atoms}/\cancel{mole}}}{(314 \times 10^{-10}\ cm)^3}$$

Step 4. As always, perform your math in a sequential manner, and simplify the expression in steps.

Perform the math in the numerator first:

$$\frac{(2\ atoms)(95.96\ grams/mole)}{6.02 \times 10^{23}\ atoms/mole} = 3.188 \times 10^{-22}\ grams$$

Perform the math in the denominator:

$$(314 \times 10^{-10}\ cm)^3 = 3.096 \times 10^{-23}\ cm^3$$

Ratio again, and solve to the final answer:

$$3.188 \times 10^{-22} \text{ grams}/3.096 \times 10^{-23} \text{ cm}^3 = 10.29 \text{ cm}^3$$

Once you get more comfortable, you can then place the entire equation sequentially into your calculator.

STUDENT PROBLEMS

1. Palladium crystalizes with a face-centered cubic system and has a density of 12.0 grams/cm³. Calculate the edge length for palladium.
   (Molar mass = 106.42 grams/mole)

2. Niobium crystallizes with a Body Center Cubic Structure and has a radius of 146 pm. What is the density of niobium? (Molar mass = 92.91 grams/mole)

   (Answers: 1. 389 pm; 2. 8.07 g/cm³)

Section 1.2 Vapor Pressure: Clausius-Clapeyron Equation

Section 1.2.1 Plotting and Linear Relationships

During this semester of chemistry, you will encounter problems with two variables measured for a system. In these cases, chemists look to see relationships between the two. For example, if you double variable A, what happens to variable B? As one variable changes, the chemist looks to see how the other variable responds. Typically, graphs are made to see if a linear relationship exists between the two variables. If a simple linear relationship exists, then the variables are denoted by the equation for a line; y = mx + b.

In order to achieve this linear relationship, mathematical operations are performed on the measured numbers. These include taking the reciprocal of a value, the log or the natural log (ln), etc. After you determine the relationship and make graphical representations, the SLOPE of this line becomes a very important quantity to solving problems seen in the second semester of chemistry.

Section 1.2.2 Pressure and Temperature – Clausius-Clapeyron Equation

The first concept and calculations discussed in regard to liquids and intermolecular forces involves vapor pressure. Because of a dynamic equilibrium between a substance's liquid and gaseous states, you can determine the pressure of the gas in this equilibrium. This pressure is referred to a substance's vapor pressure. This variable involves a temperature dependence, and as the temperature rises, the vapor pressure increases.

A plot of the natural log of pressure versus 1/temperature will yield a linear graph for which the slope multiplied by the gas constant R equals the heat of vaporization of that substance. ΔH_{vap} is the amount of energy required for the liquid phase to convert to the gaseous phase of any substance.

The Clausius-Clapeyron Equation derives from this graphical relationship:

$$\ln P = \frac{-\Delta H_{vap}}{R}\left(\frac{1}{T}\right) + \ln b$$

You can also represent this equation in a two-point form:

$$\ln \frac{P_1}{P_2} = \frac{\Delta H_{vap}}{R}\left(\frac{1}{T_2} - \frac{1}{T_1}\right)$$

You can use this equation to determine the ΔH_{vap} of a liquid or, if this quantity is known, changing pressures or temperatures.

Sample Problem 1.2 Vapor pressure and temperature – Clausius-Clapeyron Equation

The normal boiling point of ethanol is 78.4 °C. What is the vapor pressure of ethanol at 25°C? ΔH_{vap} for ethanol is 38.56 kJ/mol.

Step 1. Select the equation necessary to solve the problem. Since two temperatures are provided, use the two-point form of the Clausius-Clapeyron Equation.

$$\ln \frac{P_1}{P_2} = \frac{\Delta H_{vap}}{R} \left(\frac{1}{T_2} - \frac{1}{T_1}\right)$$

Step 2. What does the questions ask for, and what variables do you know?

Underline what the question asks for, and then circle the known quantities:

The normal boiling point of ethanol is 78.4 °C. What is the vapor pressure of ethanol at 25°C? ΔH_{vap} for ethanol is 38.56 kJ/mol.

I suggest that you use a summary table to make sure which quantities you know for the equation, especially when the problem has several variables.

Note: Temperature must be in Kelvin, and heat of vaporization and R must possess the same units, Joules or kJoules.

$T_1 = 78.4°C + 273.15 = 351.55$ P_1 = unsure

$T_2 = 25.0°C + 273.15 = 298.15$ P_2 = unknown to be solved for

$\Delta H_{vap} = 38.56$ kJ/mol

R = 8.314 J/K mol

Remember, you cannot have two unknown variables in any of these problems. Looking at this table, at first sight, P_1 and P_2 are both unknown.

Hint: The pressure at the **Normal Boiling Point** is always 1 atm. Thus, $P_1 = 1$ atm.

Step 3. Fill in the quantities, and solve.

$$\ln \frac{1}{P_2} = \frac{38560 \; J/mol}{8.314 \; J/k \; mol} \left(\frac{1}{298.15} - \frac{1}{351.55}\right) K^{-1}$$

Step 4. As always, perform your math in a sequential manner, and simplify the expression in steps.

$$\ln \frac{1}{P_2} = \frac{38560 \; \cancel{J/mol}}{8.314 \; \cancel{J}/K \; \cancel{mol}} (\frac{1}{298.15} - \frac{1}{351.55}) K^{-1}$$

$$\ln \frac{1}{P_2} = 4637.96 \; K \; (.003354 - .002845) K^{-1}$$

$$\ln \frac{1}{P_2} = 4637.96 \; K \; (.000509 \; K^{-1})$$

$$\ln \frac{1}{P_2} = 2.36072$$

You must now take the inverse natural log of both sides to simplify the equation.

$$\frac{1}{P_2} = e^{2.36072}$$

$$\frac{1}{P_2} = 10.59857$$

$$P_2 = 1/10.59857$$

$$P_2 = 0.0944 \; atm$$

Sample Problem 1.3 Vapor pressure and temperature – Clausius-Clapeyron Equation

Find the ΔH_{vap} of methylamine, which has a vapor pressure of 344 torr at -25.0°C and a boiling point of -6.4°C at 1 atm.

Step 1. Select the equation necessary to solve the problem. Since you have the two temperatures and two pressures, select the two-point form of the Clausius-Clapeyron Equation.

$$\ln \frac{P_1}{P_2} = \frac{\Delta H_{vap}}{R} (\frac{1}{T_2} - \frac{1}{T_1})$$

Step 2. What does the question ask for, and what variables do you know?

Underline what the question asks for, and then circle the known quantities:

Find the ΔH_{vap} of methylamine, which has a vapor pressure of 344 torr at -25.0°C and boiling point of -6.4°C at 1 atm.

I suggest that you use a summary table to make sure the known quantities form the equation, especially when the problem has several variables.

Note: Temperature must be in Kelvin, and heat of vaporization and R must either be both in Joules or both in kJoules.

$T_1 = -25.0°C + 273.15 = 248.15$ $P_1 = 344$ torr

$T_2 = -6.4°C + 273.15 = 266.75$ $P_2 = 1$ atm $= 760$ torr

$\Delta H_{vap} =$ UNKNOWN

$R = 8.314$ J/K mol

Step 3. Fill in the quantities, and solve.

$$\ln \frac{344 \text{ torr}}{760 \text{ torr}} = \frac{\Delta H_{vap}}{8.314 \text{ J/K mol}} \left(\frac{1}{266.75} - \frac{1}{248.15} \right) K^{-1}$$

Step 4. As always, perform your math in a sequential manner, and simplify the expression in steps.

$$\ln 0.45263 = \frac{\Delta H_{vap}}{8.314 \text{ J/K mol}} (0.003749 - 0.004029) \ K^{-1}$$

$$-0.792680 = \frac{\Delta H_{vap}}{8.314 \text{ J/K mol}} (-0.00028 \ K^{-1})$$

$$(8.314 \text{ J/K mol})(-0.792680) = \frac{\Delta H_{vap}}{8.314 \text{ J/K mol}} (-0.00028 \ K^{-1})(8.314 \text{ J/K mol})$$

$$-6.5903 \text{ J/K mol}/(-0.00028 \ K^{-1}) = \Delta H_{vap} (-0.00028)/(-0.00028)$$

$$\Delta H_{vap} = 23536.9 \text{ J/mol} = 23.5 \text{ kJ/mol}$$

STUDENT PROBLEMS

1. Octane has a vapor pressure of 40.0 torr at 45.1°C and 400 torr at 104.0°C. What is its heat of vaporization?

2. The vapor pressure of ethanol is 400.0 mmHg at 63.5°C. Its molar heat of vaporization is 39.3 kJ/mol. What is the vapor pressure of ethanol, in mmHg, at 34.9°C?

(Answers: 1. 39.0 kJ; 2. 108 mmHg)

Section 1.3 Phase Transitions: Energy Calculations

The final set of problems addressed in this chapter involves calculation of the amount of energy necessary to heat or cool a substance. You first saw the equations in the chapter that discussed the first law of thermodynamics.

The first equation depicts heat (q) or energy found when a substance changes in temperature:

$$q = \text{mass} \times \text{specific heat} \times \Delta T$$

The second equation involves the energy needed to take a substance through a phase change where the temperature does not change. The two phase changes are solid to liquid, or the melting point, and liquid to gas, or the boiling point. In these problems, a third transition is also considered: solid to gas, or sublimation. These equations for each transition are:

Solid to Liquid: $q = n \, \Delta H_{fusion}$

Liquid to Gas: $q = n \, \Delta H_{vaporization}$

Sublimation: $q = n \, \Delta H_{sublimation}$

In these equations, n = moles of substance and $\Delta H_{Sublimation} = \Delta H_{fusion} + \Delta H_{vaporization}$

Sample Problem 1.4 Energy calculations and heating curves

How much energy is needed to convert 100.0 grams of ice at -15.0°C to steam at 120.0°C.

Necessary constants for water:

$$\Delta H_{fusion} = 6.01 \text{ kJ/mol}$$

$$\Delta H_{vaporization} = 40.7 \text{ kJ/mol}$$

Specific Heat Ice = 2.09 J/g °C; Specific Heat Liquid Water = 4.18 J/g °C; Specific Heat Steam = 2.01 J/g °C

Step 1. Create a heating curve depicting the phase changes for the water. In this curve, the y-axis represents the temperature, and the x-axis represents time. Place a dot at the initial temperature (-15) and the final temperature (120).

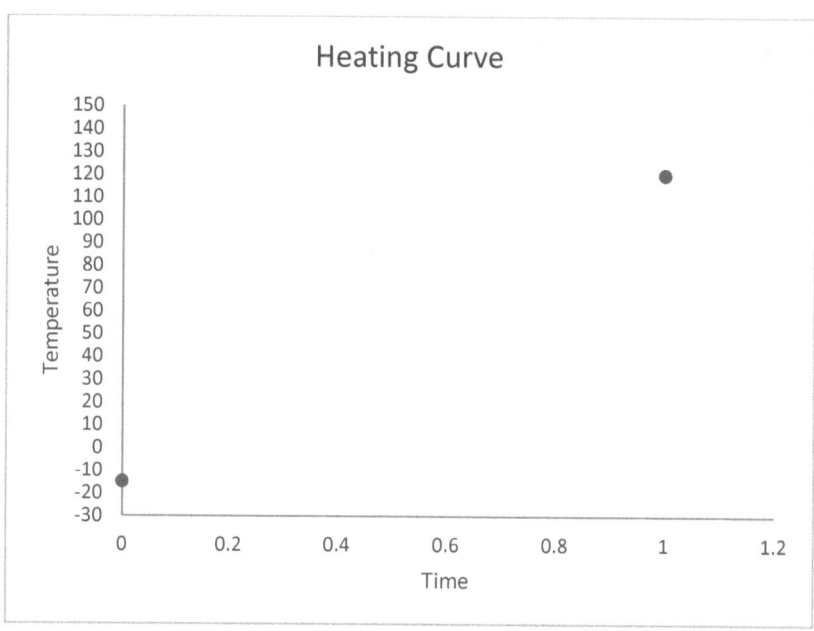

Figure 1.2a – Heating Curve – Initial Step

Step 2. First identify the temperatures where the substance melts and boils; for water, the melting point is 0°C, and the boiling point is 100°C. Place a horizontal line at each of these temperatures.

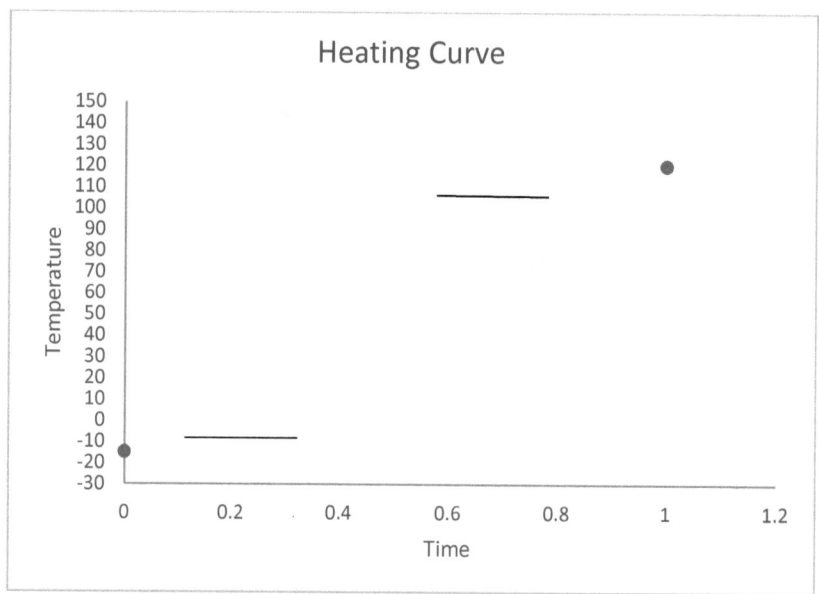

Figure 1.2b – Heating Curve – Identifying Phase Transitions

Step 3. Now, build the individual steps by connecting the dots and lines. This will correspond to the phase changes and the gains in temperature.

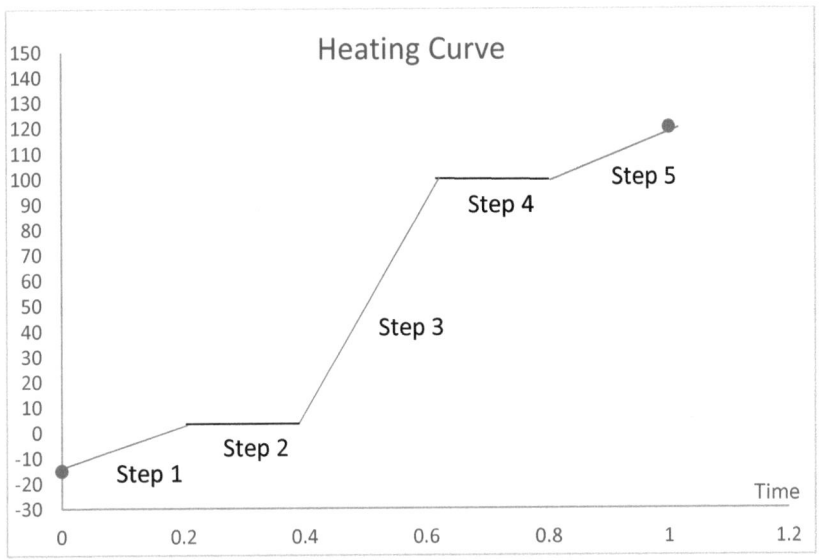

Figure 1.2c – Heating Curve – Connecting Sections

Now that step 3 is completed, you can calculate the energy involved for each step.

Step 4. Determine which equations and constants you need for each energy (q) step calculation.

For any step that rises or goes through a change in temperature, use the following equation:

$$q = \text{mass} \times \text{specific heat} \times \Delta T$$

For any step that is a horizontal line (plateau, no change in temperature), use the following equations:

$$\text{Solid to Liquid:} \quad q = n \, \Delta H_{fusion}$$

$$\text{Liquid to Gas:} \quad q = n \, \Delta H_{vaporization}$$

You will also need these constants:

$$\Delta H_{fusion} = 6.01 \text{ kJ/mol}$$

$$\Delta H_{vaporization} = 40.7 \text{ kJ/mol}$$

Specific Heat Ice = 2.09 J/g °C; Specific Heat Liquid Water = 4.18 J/g °C; Specific Heat Steam = 2.01 J/g °C

Step 5. Calculate the energy for each step, and then solve for the total energy

Calculations:

Step 1 – Rising line – 100.0 grams of ice -15°C to 0°C

Equation needed: q = mass x specific heat x ΔT

q = 100.0 grams x 2.09 J/g °C x [0 –(-15)°C]

q = 100.0 grams x 2.09 J/g °C x 15°C

q_1 = 3135 J

Step 2 – Plateau – Horizontal Line – 100.0 grams of ice melting to liquid water at 0°C

Equation Needed: q = n ΔH$_{fusion}$

ΔH$_{fusion}$ = 6.01 kJ/mol

n = grams water/Molar Mass Water = 100.0 grams/18.00 grams/mol = 5.56 moles

q = 5.56 moles x 6.01 kJ/mol

q_2 = 33.42 kJ = 33420 J

Step 3 – Rising Line – 100.0 grams Liquid Water 0°C to 100°C

Equation needed: q = mass x specific heat x ΔT

q = 100.0 grams x 4.18 J/g °C x [100 – 0°C]

q_3 = 41800 J

Step 4 - Plateau – Horizontal Line – 100.0 grams of liquid water boiling to steam at 100°C

Equation Needed: q = n ΔH$_{vaporization}$

ΔH$_{vaporization}$ = 40.7 kJ/mol

n = grams water/Molar Mass Water = 100.0 grams/18.00 grams/mol = 5.56 moles

q = 5.56 moles x 40.7 kJ/mol

q_4 = 226.3 kJ = 226300 J

Step 5 - Rising Line – 100.0 grams of steam 100°C to 120°C

Equation needed: q = mass x specific heat x ΔT

q = 100.0 grams x 2.01 J/g °C x [120 – 100°C]

$$q = 100.0 \text{ grams} \times 2.01 \text{ J/g °C} \times [20°C]$$

$$q_5 = 4020 \text{ J}$$

Total Energy: Sum the q's calculated in each step ($q_1 + q_2 + q_3 + q_4 + q_5 = q_{total}$).

$$3135 \text{ J} + 33420 \text{ J} + 41800 \text{ J} + 226300 \text{ J} + 4020 \text{ J} = 308675 \text{ J} = 309 \text{ kJ}$$

STUDENT PROBLEMS

1. Calculate the amount of energy required to turn 50.0 grams of liquid water at 35°C to steam at 180°C.

    Constants: ΔH_{fusion} = 6.01 kJ/mol $\Delta H_{vaporization}$ = 40.7 kJ/mol

    Specific Heat Ice = 2.09 J/g °C; Specific Heat Liquid Water = 4.18 J/g °C;

    Specific Heat Steam = 2.01 J/g °C

2. Calculate the amount of energy needed to vaporize 20.0 grams of acetone (C_3H_6O) at 50.0°C.

    Constants: $\Delta H_{vaporization}$ = 39.9 kJ/mol, Boiling Point = 82.3°C,

    Specific Heat Liquid Acetone = 2.44 J/g °C

(Answers: 1. 134.7 kJ; 2. 15.3 kJ)

Credits

Fig 1.1: "bravais-lattices-pif," http://biochem.co/2008/08/crystal-structure-studies/. Copyright © 2008 by Biochem.co – Biochem & Science Notes.

Chapter 2. Solutions

Up to this point, a solution has simply been explained as a solid dissolved into a liquid. Solutions can also be mixtures of both like and unlike phases or states of matter. Conceptually, this chapter in any textbook explains that solutions consist of a solute and solvent and that solutions result when like substances dissolve like substances. This chapter also explains the effect of intermolecular forces, the enthalpy of solution formation, and the factors (temperature and pressure) that affect solubility. Problems in this chapter involve the calculation of Henry's Law, concentration, and colligative properties.

Section 2.1 Henry's Law

The first type of problem encountered in this chapter involves Henry's Law and the effect of pressure on the solubility of gases. Simply, as the pressure over a liquid increases, the concentration of a gas dissolved in a liquid increases. In other words, the solubility of the gas is directly proportional to the partial pressure of the gas above the liquid.

The equation used is:

$$\text{Solubility} = \text{Henry's Law constant} \times \text{Pressure}$$

$$S = k\,P$$

Solubility is expressed as Molarity = M, k = M/atm and Pressure = atm

Sample Problem 2.1 Henry's Law

What quantity of carbon dioxide, CO_2, will dissolve in a soda pop at a pressure of 3.0 atm?

k for CO_2 = 3.4 x 10^{-2} M/atm

Step 1. Use Henry's Law equation, fill in the correct variables, and solve for the unknown.

$$\text{Henry's Law: } S = k\,P$$

$$S = k\,P$$

Solubility is the unknown, and the question provides k and P.

$$S = 3.4 \times 10^{-2} \text{ M/atm} \times 3.0 \text{ atm}$$

$$S = 0.10 \text{ M}$$

STUDENT PROBLEMS

1. What pressure do you need to dissolve 0.050 M NH_3 in water? NH_3 k = 58 M/atm
2. How much O_2 is dissolved in water at a pressure of 2.5 atm? O_2 k = 1.3 x 10^{-3} M/atm

(Answers: 1. 0.00086 atm; 2. 0.0032 M)

Section 2.2 Concentration Expressions and Calculations

There are numerous ways to express the concentration of a solute in a solvent. Remember that, in all cases, the solute is the minority phase (by mass), the solvent is the majority phase, and the solution is the solute + solvent.

Concentration is the amount of solute dissolved in a solvent.

Section 2.2.1 Molarity (M)

Molarity is expressed as:

$$M = \frac{\text{moles of solute}}{\text{L of solution}}$$

Sample Problem 2.2 Molarity

What is the Molarity of a 100.0 L solution in which 99.0 grams of NaCl has dissolved?

Step 1. Identify the correct equation, fill in the known variables, and solve for unknown.

$$M = \frac{\text{moles of solute}}{\text{L of solution}}$$

You do not know the molarity, you do know 100.0 L of a solution, and you do know 99.0 grams of NaCl.

NaCl is the solute and must appear in units of moles.

Remember:

$$\text{Moles}_{NaCl} = \text{grams}_{NaCl}/\text{Molar Mass}_{NaCl} = 99.0 \text{ grams NaCl}/58.50 \text{ grams/mol}$$

$$\text{Moles}_{NaCl} = 1.69 \text{ moles NaCl}$$

Step 2. Solve for M.

$$M = \frac{1.69 \text{ moles}}{100.0 L} = 0.0169 \text{ M}$$

STUDENT PROBLEMS

1. How much solution do you need for a 1.88 M solution that has 2.33 moles of K_2SO_4?
2. How many grams of $LiNO_3$ do you need to prepare 500.0 mL of a 0.087M solution?
(Answers: 1. 1.24 L; 2. 3.0 grams)

Section 2.2.2 Molality

Molality is expressed as:

$$m = \frac{\text{moles of solute}}{\text{kg of solvent}}$$

Sample Problem 2.3 Molality

What is the molality of a solution prepared by dissolving 25.0 grams of CsBr in 500.0 grams of water?

Step 1. Select the correct equation, identify the unknown quantity, and determine the known variables.

The problem asks for molality:

$$m = \frac{\text{moles of solute}}{\text{kg of solvent}}$$

Molality stands as the unknown in the problem, and you know the moles of solute and kg of solvent.

Step 2. Fill in the quantities, and solve.

$$m = \frac{\text{moles of solute}}{\text{kg of solvent}}$$

moles of solute = grams of CsBr/Molar Mass$_{CsBr}$ = 25.0 grams/212.81 grams/mole = 0.117 moles

kg of solvent = kg of water = 500.0 grams/1000 grams/kg = 0.500 kg water

$$m = \frac{0.117 \text{ moles}}{0.500 \text{ kg}} = 0.235 \text{ m}$$

STUDENT PROBLEMS

1. What is the molality of 10.0 grams of $FeCl_3$ dissolved in 800.0 mL of water at 25.0°C?
2. How many kg of water do you need to dissolve 88.8 moles of $CaSO_3$ to create a 4.44 molal solution?

(Answers: 1. 0.0771 m 2. 20.0 kg)

Section 2.2.3 Percent by Mass, ppm, ppb

Percent by mass for either the solute or solvent is expressed as:

$$\% \text{ mass}_{solute} = \frac{\text{mass of solute}}{\text{mass of solution}} \times 100\%$$

$$\% \text{ mass}_{solvent} = \frac{mass\ of\ solvent}{mass\ of\ solution} \times 100\%$$

Two other units that express the mass amount of the solute with respect to the mass of the solution are parts per million (ppm) and parts per billion (ppb). By equation, both of these units are expressed as:

$$ppm = \frac{mass\ of\ solute}{mass\ of\ solution} \times 10^6$$

$$ppb = \frac{mass\ of\ solvent}{mass\ of\ solution} \times 10^9$$

Sample Problem 2.4 Percent by mass

How many grams of HCl do you need to prepare 150.0 grams of a 12.2% solution of HCl?

Step 1. Select the correct equation, identify the unknown quantity, and determine the known variables.

$$\% \text{ mass}_{solute} = \frac{mass\ of\ solute}{mass\ of\ solution} \times 100\%$$

You do not know the grams of solute, HCl, but you do know the % by mass and mass of solution.

Step 2. Fill in the quantities, and solve.

$$\% \text{ mass}_{solute} = \frac{mass\ of\ solute}{mass\ of\ solution} \times 100\%$$

$$12.2\% = \frac{mass\ of\ solute}{150.0\ grams} \times 100\%$$

$$0.122 = \frac{mass\ of\ solute}{150.0\ grams}$$

Mass of solute = 0.122 x 150.0

Mass of solute = 18.3 grams HCl

STUDENT PROBLEMS

1. How many ppm of NaI are present in a 0.001% NaI solution?
2. How many grams of water do you need to prepare 500.0 grams of a 0.05% $Pb(NO_3)_2$ solution?

(Answers: 1. 10 ppm; 2. 499.75 grams)

Section 2.2.4 Mole Fraction

The mole fraction for either the solute or solvent is expressed by:

$$X_{solute} = \frac{moles\ of\ solute}{moles\ of\ solution}$$

$$X_{solvent} = \frac{moles\ of\ solvent}{moles\ of\ solution}$$

Sample Problem 2.5 Mole Fraction

What are the mole fractions of glucose, $C_6H_{12}O_6$, and water that contains 8.88 grams of glucose in 333.3 grams of water?

Step 1. Select the correct equation, identify the unknown quantity, and determine the known variables.

$$X_{solute} = \frac{moles\ of\ solute}{moles\ of\ solution}$$

You do not know the mole fraction, but you do know the grams of both the solute and solvent.

Since the problem asks for moles, you must convert the grams of each component to moles by:

$$Moles = grams/Molar\ Mass$$

Step 2. Convert grams to moles, fill in the quantities, and solve.

Moles of glucose = 8.88 grams/180.00 grams/mole = 0.0493 moles

Moles of water = 333.3 grams/18.00 grams/mole = 18.52 moles

Total moles of solution = 18.52 moles glucose + 0.0493 moles water = 18.57 moles total

$$X_{solute} = \frac{moles\ of\ solute}{moles\ of\ solution}$$

$$X_{glucose} = \frac{0.0493\ moles}{18.57\ moles} = 0.00265$$

With a 2-component solution, the sum of the individual mole fractions will add up to 1. Therefore, the mole fraction of the water is:

$$X_{water} = 1 - X_{glucose}$$

$$X_{water} = 1 - 0.00265$$

$$X_{water} = 0.99735$$

STUDENT PROBLEMS

1. What is the mole fraction of ethanol, CH_3CH_2OH, in a solution prepared by dissolving 55.5 grams of ethanol in 135.7 grams of water?
2. How many moles of water do you need to prepare a solution of acetic acid, $HC_2H_3O_2$, that contains a 0.15 mole fraction of acetic acid?

(answers: 1. 0.138; 2. 0.85 moles)

Section 2.2.5 Converting between Concentration Units

Conversions between concentration units are very common problems in the solutions chapter of any general chemistry textbook. The key conversion factor involves the density of the solution, which allows for the determination of either the grams or volume of the solution.

Sample Problem 2.6 Combined molarity, molality, % by mass and mole fraction calculations

What is the molality, molarity, and mole fraction of each component in a 25.0% solution of HNO_3? The density of the solution is 1.23 grams/mL.

Step 1. Select the correct equations, identify the unknown quantity, and determine the known variables.

You know the percent by mass and the density of the solution; you do not know the molarity, molality and mole fraction. This problem asks you to solve for three different quantities, and it is suggested you solve them individually step by step.

Step 2. Identify the solute and the solvent.

There is a 25.0% solution of HNO_3, which means that the HNO_3 is the solute, and water is the solvent.

Step 3. Solve for molality, m.

$$\underline{m} = \frac{\text{moles of solute}}{\text{kg of solvent}}$$

Moles of Solute = moles of HNO_3 = grams HNO_3/Molar Mass HNO_3

The grams of HNO_3 come from the percent by mass value, 25.0%. This means that there exists 25.0% of HNO_3 and 75.0% of water in this solution, or 25.0 grams of HNO_3 and 75.0 grams of water.

Moles of HNO₃ = 25.0 grams/63.00 grams/mole = 0.397 moles HNO₃

kg of solvent = kg of water = 0.075 kg water

$$m = \frac{\text{moles of solute}}{\text{kg of solvent}}$$

$$m = \frac{0.397 \text{ moles}}{0.075 \text{ kg}} = 5.29 \text{ m}$$

Step 4. Solve for molarity.

$$M = \frac{\text{moles of solute}}{\text{L of solution}}$$

For this step, you need moles of HNO₃ and the L of the solution. You solve for moles of HNO₃ in Step 3, which came to 0.397 moles HNO₃. The L of solution requires the density provided in the question, 1.23 gram/mL. Since you know % by mass, and there are 25.0 grams of HNO₃ and 75.0 grams of water, 100.0 grams of solution exist (25.0 + 75.0).

L of solution

$$\text{mL of solution} = \frac{\text{grams of solution}}{\text{density of solution}}$$

$$\text{mL of solution} = \frac{100.0 \text{ grams}}{1.23 \text{ grams/mL}} = 81.30 \text{ mL}$$

L of solution = mL of solution/1000 mL/L = 81.30 mL/1000 mL/L = 0.08130 L

Solving for M

$$M = \frac{\text{moles of solute}}{\text{L of solution}}$$

$$M = \frac{0.397 \text{ moles}}{0.08130 \text{ L}} = 4.8 \text{ M}$$

Step 5. Solve for mole fraction.

$$X_{\text{solute HNO3}} = \frac{\text{moles of solute}}{\text{moles of solution}}$$

Moles of HNO₃ are known = 0.397 moles

Total moles of solution = moles HNO₃ + moles water

Moles of water = grams water/molar Mass water = 75.0 grams/18.00 grams/mole = 4.17 moles

Total moles = 0.397 + 4.17 = 4.567 moles

$$X_{solute\ HNO3} = \frac{0.397\ moles}{4.567\ moles} = 0.087$$

$$X_{water} = 1 - 0.087 = 0.913$$

STUDENT PROBLEMS

1. What are the molality, molarity, and mole fraction of each component in a 15.0% solution of H_2SO_4? The density of the solution is 1.05 grams/mL.
2. What is the percent by mass and molality of a 2.33 M solution of LiBr. The density of this solution is 1.19 grams/mL.

   (Answers: 1. 1.80 m, 1.61 M, X_{water} = 0.997, X_{H2SO4} = 0.003;

   2. 2.36 m, 17.0% LiBr, 83.0% Water)

Section 2.3 Colligative Properties

A colligative property is a property of a solution that is affected by the concentration of solute molecules or solute ions and not by the chemical makeup of that solute. The four properties addressed in these problems include vapor pressure, freezing point, boiling point and osmotic pressure.

Section 2.3.1 Vapor Pressure Lowering

When you add a solute to a solvent, the vapor pressure of that solvent lowers because of increased intermolecular forces now created in solution. In these problems, the solute can be either volatile or non-volatile. Thus, you will use two equations to calculate this property according to Raoult's Law.

Section 2.3.1.1 Non-Volatile Solute

If the solute is non-volatile (e.g., a salt or sugar), the equation used to find the new pressure is:

$$P_{solvent} = \chi_{solvent} \, P°_{solvent}$$

In many of these problems, the subscripts can be the number 1 or letter A, which represent the solvent or the subscripts can be a 2 or a B which represent the solute.

You can use the other equation for this scenario to determine the change in the vapor pressure:

$$\Delta P = \chi_{solute} \, P°_{solvent}$$

Sample Problem 2.7 Colligative Properties – Vapor Pressure

Calculate the vapor pressure of a solution containing 10.5 grams of $MgCl_2$ in 100 grams of water at 30.0°C. The pure vapor pressure of water at this temperature is 31.8 torr.

Step 1. Select the correct equations, identify the unknown quantity and determine the known variables. Also, select the solute and solvent.

Calculate the vapor pressure of a solution containing 10.5 grams of $MgCl_2$ in 100.0 grams of water at 30.0°C. The pure vapor pressure of water at this temperature is 31.8 torr.

In this problem, water serves as the solvent (greater mass quantity, and its pure vapor pressure is given), and the $MgCl_2$ serves as the solute.

You do not know the new vapor pressure; you do know the pure vapor pressure; you must calculate the mole fraction of the solvent, water, from the gram quantities provided.

Equation:

$$P_{solvent} = \chi_{solvent} P^o_{solvent}$$

Step 2. Calculate the mole fraction of water.

$$\chi_{solvent\ water} = \frac{moles\ of\ water}{moles\ of\ solution}$$

Moles of water = grams water/Molar Mass Water = 100.0 grams/18.00 grams/mole = 5.56 moles water

Moles of solution = moles water + moles $MgCl_2$

Moles $MgCl_2$ = grams $MgCl_2$/Molar Mass $MgCl_2$ = 10.5 grams/95.21 grams/mole = 0.110 moles $MgCl_2$

Moles of solution = 0.110 + 5.56 = 5.67 moles

$$\chi_{solvent\ water} = \frac{moles\ of\ water}{moles\ of\ solution}$$

$$\chi_{solvent\ water} = \frac{5.56\ moles}{5.67\ moles} = 0.98$$

Step 3. Calculate the new vapor pressure.

$$P_{solvent} = \chi_{solvent} P^o_{solvent}$$

$$P_{solvent} = 0.98 \times 31.8\ torr$$

$$P_{solvent} = 31.2\ torr$$

Answer check: Remember the vapor pressure of a liquid lowers when you add a solute.

Section 2.3.1.2 Volatile Solute

In vapor pressure problems, the solute can be a volatile substance. Volatile solutions will be mixtures of organic compounds, such as benzene, toluene, acetone, octane, pentane, hexane, etc. The vapor pressure of the solution, if behaving ideally, will follow Raoult's Law, and the equation used to solve these types of problems for a two-component mixture is:

$$P_{total} = \chi_1 P^o_1 + \chi_2 P^o_2$$

Where you add the mole fraction of each component times its pure vapor pressure together, which equals the total pressure exerted by the solution.

Sample Problem 2.8 Colligative properties – vapor pressure and volatile solutes

A solution contains 10.2 grams of hexane (C_6H_{14}) and 15.6 grams of benzene (C_6H_6). What is the vapor pressure of the solution? The vapor pressure of pure hexane is 151 torr, and pure benzene is 75 torr at room temperature.

Step 1. Select the correct equations, identify the unknown quantity, and determine the known variables.

A solution contains 10.2 grams of hexane (C_6H_{14}) and 15.6 grams of benzene (C_6H_6). What is the vapor pressure of the solution? The vapor pressure of pure hexane is 151 torr, and pure benzene is 75 torr at room temperature.

You have the amounts and pure pressures of the two components of the mixture; you do not know the total pressure. Both of these substances are volatile compounds; thus, the equation you need to solve is:

$$P_{total} = \chi_1 P_1^o + \chi_2 P_2^o$$

Step 2. Calculate mole fractions of each component.

Since the problem gives the gram quantities, you must convert each to moles to achieve the calculation of the mole fraction. Since the sum of the individual moles fractions equals one, you only have to determine one substance.

Moles Hexane = grams hexane/molar mass Hexane = 10.2 grams/86.00 grams/mole = 0.119

Moles Benzene = grams benzene/molar mass Benzene = 15.6 grams/78.00 grams/mole = 0.200

Total Moles = 0.119 + 0.200 = 0.319

$$\chi_{hexane} = 0.119/0.319 = 0.373 \qquad \chi_{benzene} = 1 - 0.373 = 0.627$$

Step 3. Calculate the total pressure.

$$P_{total} = \chi_1 P_1^o + \chi_2 P_2^o$$

$$P_{total} = (0.373 \times 151 \text{ torr}) + (0.627 \times 75 \text{ torr})$$

$$P_{total} = 103 \text{ torr}$$

STUDENT PROBLEMS

1. At 25.0°C, the vapor pressure of pure water is 23.8 mmHg. What is the change in pressure when 13.4 grams of NaCl is added to 500.0 mL of water?

2. What mole fraction of sucrose, $C_{12}H_{22}O_{11}$, lowers the vapor pressure of water at 25.0°C to 21.6 torr? The pure vapor pressure of water is 23.8 torr.

3. A solution contains 6.88 grams of heptane (C_7H_{16}) and 6.88 grams of octane (C_8H_{18}). What is the vapor pressure of the solution? The vapor pressure of pure heptane is 45.8 torr, and pure octane is 10.9 torr at room temperature.

(Answers: 1. ΔP = 0.195 mmHg, 2. $X_{sucrose}$ = 0.0924; 3. 29.5 torr)

Section 2.3.2 Boiling Point Elevation and Freezing Point Lowering

You will solve these two colligative properties together, because the equations are nearly alike. When you add a solute to a solvent, the intermolecular forces increase, which leads to an increase in the temperature at which the solvent boils or a decrease in the temperature at which it will freeze. The two equations used to solve these types of problems are:

$$\Delta T_b = K_b\, m$$

$$\Delta T_f = K_f\, m$$

Where K_b and K_f are the boiling point elevation and freezing point depression constants and m is molality

Sample Problem 2.9 Colligative properties – boiling point and freezing point

Add 3.00 grams of ethylene glycol or antifreeze, $C_2H_6O_2$, to 100.0 grams of water. What will the freezing temperature be for this mixture? At which temperature will it boil?
K_f of water is 1.858 °C/m and K_b = 0.512 °C/m

Step 1. Select the correct equations, identify the unknown quantity, and determine the known variables.

Add 3.00 grams of ethylene glycol or antifreeze, $C_2H_6O_2$, is added to 100.0 grams of water. What will the freezing temperature be for this mixture? At which temperature will it boil?
K_f of water is 1.858 °C/m

Solving for the freezing point first, the new freezing point is unknown, and the amount of water and ethylene glycol are known.

The equation required is: $\Delta T_f = K_f m$

Step 2. Determine the solvent, solute, and molality, m.

Since the problem gives K_f and asks for ΔT, the only other variable that you must determine before the final answer is the molality, m.

$$m = \frac{moles\ of\ solute}{kg\ of\ solvent}$$

In this problem, water serves as the solvent (greater mass and K_f is provided), and ethylene glycol serves as the solute.

Moles of solute = grams$_{ethylene\ glycol}$/molar mass$_{ethylene\ glycol}$ = 3.00 grams/62.00 grams/mole = 0.0484 moles

kg of solvent = kg of water = 100.0 grams/1000 grams/kg = 0.100 kg water

$$m = \frac{moles\ of\ solute}{kg\ of\ solvent}$$

$$m = \frac{0.0484\ moles}{0.100\ kg}$$

m = 0.484 m

Step 3. Solve for the change in temperature, and calculate the new freezing point.

$$\Delta T_f = K_f\ m$$

$$\Delta T_f = (1.858\ °C/m)(0.484\ m)$$

$$\Delta T_f = 0.899\ °C$$

New Freezing Temperature = 0°C - ΔT = 0°C − 0.899°C = − 0.899°C

Step 4. Solve for the new boiling point using the calculated molality, 0.484 m, and K_b value.

$$\Delta T_b = K_b\ m$$

$$\Delta T_b = (0.512\ °C/m)(0.484\ m)$$

$$\Delta T_b = 0.248\ °C$$

New Boiling Temperature = 100°C + ΔT = 100.000°C + 0.248°C = 100.248°C

STUDENT PROBLEMS

1. What is the molality of an aqueous solution that has a freezing point of -1.6 °C?
2. Calculate the freezing and boiling points of a solution in which 22.2 grams of sucrose, $C_{12}H_{22}O_{11}$, dissolved in 222.2 grams of water?

(Answers: 1. 0.86 m; 2. Freezing Point = -0.543 °C, Boiling Point = 100.15 °C)

Section 2.3.3 Osmotic Pressure

The final colligative property is for the calculation of osmotic pressure which is the pressure, when applied to a solution, that stops osmosis. The equation that you will use to determine this pressure is:

$$\Pi = MRT$$

Where Π is the osmotic pressure in units of atm, M is molarity, T is temperature in Kelvin, and R is the ideal gas law constant 0.08206 L atm/mole K

Sample Problem 2.10 Colligative properties – osmotic pressure

A 350.0 mL solution of ethylene glycol in water has an osmotic pressure of 6.78 mmHg at 25°C. How many grams of ethylene glycol are in this solution?

Step 1. Select the correct equation, identify the unknown quantity, and determine the known variables.

A 350.0 mL solution of ethylene glycol in water has an osmotic pressure of 6.78 mmHg at 25.0°C. How many grams of ethylene glycol are in this solution?

You do not know the amount of ethylene glycol, while you do know the temperature, mL of solution, and the osmotic pressure. Thus, the needed equation is:

$$\Pi = MRT$$

Step 2. Determine molarity, M.

Since the amount of ethylene glycol, the solute, is part of the M equation, solve for M from the other variables. Make sure all quantities are in the correct units.

$$M = \frac{moles\ of\ solute}{L\ of\ solution}$$

Osmotic pressure in atm = 6.78 mmHg/760 mmHg/atm = 0.0089 atm

R = 0.08206 L atm/mole K

T in Kelvin = °C + 273.15 = 25.0°C + 273.15 = 298.15

Solve for M now that each variable is in the correct units:

$$\Pi = M R T$$

0.0089 atm = M x 0.08206 L atm/mol K x 298.15 K

$$M = \frac{0.0089 \text{ atm}}{0.08206 \text{ L atm/mol K} \times 298.15 \text{ K}}$$

M = 0.000364

Step 3. Determine the amount of ethylene glycol, $C_2H_6O_2$, from the M.

$$\underline{M} = \frac{moles\ of\ solute}{L\ of\ solution}$$

The L of the solution are known, 350.0 mL or 0.350 L, and you solved for the M in step 2, M =0.000364. Solving the equation for moles of ethylene glycol:

$$\underline{M} = \frac{moles\ of\ solute}{L\ of\ solution}$$

M x L$_{solution}$ = moles ethylene glycol

Moles ethylene glycol = 0.000364 moles/L x 0.350 L = 0.000127 moles

Determine grams of ethylene glycol:

Grams$_{ethylene\ glycol}$ = moles$_{ethylene\ glycol}$ x molar mass$_{ethylene\ glycol}$

= 0.000127 x 62.00 grams/mole = 0.00789 grams ethylene glycol

STUDENT PROBLEMS

1. What is the osmotic pressure of a 0.500 M solution of sucrose at 35.0°C?
2. 18.88 grams of glycerol are dissolved in water to make a 1200.0 mL solution. What is the osmotic pressure of this solution at 55.5°C?

(Answers: 1. 12.6 atm; 2. 4.61 atm)

Section 2.3.4 Colligative Properties and Ionic Solutions, van 't Hoff Factor

All previous colligative property problems had a non-ionic solute. In the next set of problems, an ionic solute will be dissolved, thus leading to the inclusion of an additional variable i, the van 't

Hoff factor, to the equations for vapor pressure, osmotic pressure, boiling point elevation, and freezing point depression.

$$\Pi = i\,M\,R\,T$$

$$\Delta T_b = i\,K_b\,m$$

$$\Delta T_f = i\,K_f\,m$$

$$\Delta P = i\,\chi_{solute}\,P^{\circ}_{solvent}$$

Section 2.3.4.1 van 't Hoff Factor Determination

The van 't Hoff factor can be determined using these equations, or you can simply determine it by counting the number of ions present when the ionic compound dissolves in water. For example,

NaCl in solution splits to one ion of Na$^+$ and one ion of Cl$^-$; thus, the van 't Hoff factor is 2 (1 +1).

AlCl$_3$ in solution splits to one ion of Al^{+3} and 3 ions of Cl$^-$; thus, the van 't Hoff factor is 4 (3 +1).

Sample Problem 2.11 Colligative properties – van 't Hoff factor

What is the freezing point of a solution that contains 25.0 grams of NaCl dissolved in 1000 grams of water? K_f of water is 1.858 °C/m

Step 1. Select the correct equation, identify the unknown quantity, and determine the known variables.

These types of problems are solved the exact same way as described in Section 2.3, but they now include the van 't Hoff factor.

What is the freezing point of a solution that contains 25.0 grams of NaCl dissolved in 1000 grams of water? K_f of water is 1.858 °C/m

The freezing point is unknown, and the amounts of NaCl and water are known.

Step 2. Determine the molality and van 't Hoff factor.

van 't Hoff factor for NaCl = 2 (one sodium and one chlorine ion)

Molality is:

$$\underline{m} = \frac{moles\ of\ solute}{kg\ of\ solvent}$$

Moles of solute, NaCl = grams NaCl/molar mass NaCl = 25.0 grams/ 58.45 grams/mole = 0.428 moles

kg of solvent, water = 1000 grams/1000 grams/kgram = 1 kg water

$$m = \frac{0.428\ moles}{1\ kg}$$

m = 0.428 m

Step 3. Solve for temperature change and new freezing point.

$$\Delta T_f = i\ K_f\ m$$

$$\Delta T_f = 2 \times 1.858\ °C/m \times 0.428\ m$$

$$\Delta T_f = 1.59\ °C$$

Freezing Point = 0°C − ΔT = 0°C − 1.59°C = −1.59°C

STUDENT PROBLEMS

1. What is the osmotic pressure of a 0.670 M MgF_2 solution at 70.0°C?
2. What is the change in vapor pressure of 10.5 grams of Li_2SO_4 in 250.0 grams of water at 25.0°C? Pure vapor pressure of water is 23.8 torr.
3. Calculate the van 't Hoff Factor for a 3.33 molal solution of K_3PO_4 with a boiling point of 102.0°C?

(Answers: 1. 56.6 atm; 2. ΔP = 0.163 torr; 3. i = 1.17)

Section 2.3.5 Additional Colligative Property Problems, Calculation of Molar Mass

These sets of problems can prove a little more complex when you must determine the molar mass of the solute. This section presents the simplistic method to solve the more difficult problems for any colligative property.

Sample Problem 2.12 Colligative properties – molar mass calculations

What is the molar mass of 11.3 grams of an unknown solute dissolved in 250.0 grams of water with a solution boiling point of 101.55°C? K_b for water is 0.512 °C/m

Step 1. Select the correct equation, identify the unknown quantity, and determine the known variables.

What is the molar mass of 11.3 grams of an unknown solute dissolved in 250.0 grams of water with a solution boiling point of 101.55°C? K_b for water is 0.512 °C/m

The boiling point elevation equation is needed:

$$\Delta T_b = K_b \, m$$

The molar mass is the unknown variable, which is part of the molality value, thus the change in temperature needs to be determined.

Step 2. Determine ΔT, and solve for molality, m.

ΔT = 101.55 − 100 = 1.55 °C

You now have the molality:

$$\Delta T_b = K_b \, m$$

$$1.55 \text{ °C} = 0.512 \text{ °C/m} \times m$$

$$m = 3.03 \text{ m}$$

Step 3. Determine moles of unknown from molality.

Molality, m is defined by

$$m = \frac{\text{moles of solute}}{\text{kg of solvent}}$$

Solve for moles with the molality above and kg of water in the problem

$$3.03 \text{ m} = \frac{\text{moles of solute}}{0.250 \text{ kg}}$$

Moles of unknown solute = 3.03 moles/kg × .250 kg = 0.758 moles of unknown

Step 4. Determine the molar mass.

Molar mass is grams/moles. Simply, the molar mass of the unknown is the 11.3 grams divided by 0.758 moles:

$$\text{Molar Mass Unknown} = \frac{11.3 \text{ grams}}{0.758 \text{ moles}} = 14.91 \text{ grams/mole}$$

Another way of solving these types of problems involves looking at the equations with the inclusion of the molar mass variable. Molarity and molality can be expressed as:

$$m = \frac{\text{moles of solute}}{\text{kg of solvent}} = \frac{\text{grams/MM}}{\text{kg of solvent}}$$

$$M = \frac{\text{moles of solute}}{\text{L of solution}} = \frac{\text{grams}/MM}{\text{L of solution}}$$

STUDENT PROBLEMS

1. What is the molar mass of 20.8 grams of an unknown solute for which a 350.0 mL solution possesses an osmotic pressure of 15.0 mmHg at 50.0°C?
2. What is the molar mass of 13.3 grams of an unknown solute dissolved in 0.55 kg of water that exhibits a freezing temperature of -0.88°C?

(Answers: 1. 8.0 x 10^4 grams/mole; 2. 51.06 grams/mole)

Chapter 3. Kinetics

Section 3.1 Determining the Rate of a Reaction via Stoichiometry

Because the mole amounts of either products or reactants can be related by stoichiometry, the rates by which the products form or by which the reactants are consumed can also be related. Remember, sign convention can prove key in these problems; negatives are associated with reactants, which deplete, and positives are associated with products, which form.

A stoichiometric type of kinetic problems is typically identified by only having the statement of a rate for a component as it relates to the rate of another component in that given reaction. Also, recall that, for any equation in which you will determine the rate, you will use the reciprocal of the coefficient of that component in the chemical reaction. For example, in this reaction:

$$3A + 5B \rightarrow 2C$$

$$\frac{1}{3} \text{Rate}_A = \frac{1}{5} \text{Rate}_B = \frac{1}{2} \text{Rate}_C$$

Where A and B have negative values, because each are reactants, and C possesses a positive value, because it is the product.

Sample Problem 3.1 Rate determination via stoichiometry

For the reaction:

$$H_2(g) + I_2(g) \rightarrow 2HI(g)$$

If the rate of H_2 consumption is 1.60×10^{-3} M/sec, what is the rate of formation of HI?

Step 1. Identify the type of Kinetic Problem, the known and unknown variables in the problem.

You can identify a stoichiometric type of kinetic problems by it having only the statement of a rate for a component as it relates to the rate of another component in that given reaction.

Thus, for the reaction:

$$H_2(g) + I_2(g) \rightarrow 2HI(g)$$

If the rate of H_2 consumption is 1.60×10^{-3} M/sec, what is the rate of formation of HI?

You know the Rate of H_2, and you do not know the rate of HI.

Step 2. Determine the stoichiometric ratio, and solve.

For this reaction, the rate of HI forms at twice the rate of H_2 depletion, or:

$$\frac{1}{1}\text{Rate}_{H2} = \frac{1}{2}\text{Rate}_{HI}$$

Or

$$2\,\text{Rate}_{H2} = \text{Rate}_{HI}$$

$$2\,(1.60 \times 10^{-3}\text{ M/sec}) = \text{Rate}_{HI}$$

$$\text{Rate}_{HI} = 3.2 \times 10^{-3}\text{ M/sec}$$

STUDENT PROBLEMS

1. For the reaction, $2\,H_2O_2\,(aq) \rightarrow 2\,H_2O\,(l) + O_2\,(g)$, the rate of formation of the water is 3.33×10^{-2} M/sec. What is the rate of formation of the O_2? What is the rate of depletion of the peroxide?
2. For the reaction, $C_3H_8\,(g) + 5\,O_2\,(g) \rightarrow 3\,CO_2\,(g) + 4\,H_2O\,(g)$, the rate of formation of water is 2.86×10^{-4} M/sec. What are the rates for all the other components in this reaction?

(Answers: 1. Rate O_2 = 1.67×10^{-2} M/sec, Rate H_2O_2 = 3.33×10^{-2} M/sec; 2. Rate CO_2 = 2.15×10^{-4} M/sec, Rate O_2 = 3.58×10^{-4} M/sec, Rate C_3H_8 = 7.15×10^{-5} M/sec)

Section 3.2 Dependence of Rate on Concentration; Rate Law Determination

Given the reaction 3 A + 2 B → 2 C, the rate law expression depends on the reactants and is written initially as:

$$\text{Rate} = k[A]^a[B]^b$$

Where k represents the rate constant for the reaction, and the lower-case a and b represent the orders for each of the reactants.

In these problems, you must utilize experimental data containing concentration and rates to solve for the orders and k so that you can write the complete rate law.

Sample Problem 3.2 Rate law determination

Determine the rate law for:

$$C_2H_4\ (g) + I_2\ (g) \rightarrow C_2H_4I_2\ (g)$$

Given the following experimental data:

Experiment	$[C_2H_4]$ (M)	$[I_2]$ (M)	Rate (M/sec)
1	0.01 M	0.02 M	0.022
2	0.05 M	0.04 M	0.048
3	0.01 M	0.08 M	0.088
4	0.02 M	0.02 M	0.088

Step 1. Identify the type of kinetic problem and the known and unknown variables in the problem.

Since a chart of data with concentrations and rates is provided, this is a rate law problem. The unknown variables are the orders for both I_2 and C_2H_4.

Step 2. Create a simple expression for the rate law, determination of orders.

The first expression of the rate law for this reaction is:

$$\text{Rate} = k[C_2H_4]^x[I_2]^y$$

You must determine the orders x and y followed by the rate constant, k.

Step 3. Determine the order of x.

All this means is what numerical value does x have for this reaction. Here, the table of data comes into use. To solve for x or how C_2H_4 concentration effects the rate of a reaction, two lines of data that must be selected where the concentration of the other reactant, I_2, is held constant.

Experiment	[C_2H_4] (M)	[I_2] (M)	Rate (M/sec)
1	0.01 M	0.02 M	0.022
2	0.05 M	0.04 M	0.048
3	0.01 M	0.08 M	0.088
4	0.02 M	0.02 M	0.088

Same concentration of I_2

So, you will use Experiments 1 and 4 to solve for x because the iodine concentration is 0.02 M in each experiment. x is determined by taking the ratio of the values in experiment 4 to that of 1 or taking the ratio of experiment 1 to 4.

To solve for any order, you can use this expression:

$$\left(\frac{Concentration_{Exp\ A}}{Concentration_{Exp\ B}}\right)^x = \left(\frac{Rate_{Exp\ A}}{Rate_{Exp\ B}}\right)$$

Where x is the order. Thus, ratio the values in experiment 4 to 1.

$$\left(\frac{0.02}{0.01}\right)^x = \left(\frac{0.088}{0.022}\right)$$

$$(2)^x = (4)$$

$$x = 2$$

For this expression, you can easily see that 2 raised to the 2nd power equals 4. Many times in problems, the math does not appear as readily. In this case, you will run a log through the expression:

$$(2)^x = (4)$$

$$\text{Log}\ (2)^x = \log\ (4)$$

$$x \log 2 = \log 4$$

$$x = \log 4 / \log 2$$

$$x = 2$$

The rate law now goes to:

$$\text{Rate} = k\,[C_2H_4]^2\,[I_2]^y$$

And y must be determined.

Step 4. Determine the order of y.

You will solve the order of y in the exact same manner as you did for x. The only difference is that the problem now looks at how I_2 concentration effects the rate; therefore, the concentration of C_2H_4 must be held constant. From the data table:

Experiment	[C₂H₄] (M)	[I₂] (M)	Rate (M/sec)
1	0.01 M	0.02 M	0.022
2	0.05 M	0.04 M	0.048
3	0.01 M	0.08 M	0.088
4	0.02 M	0.02 M	0.088

Experiments 1 and 3 show that the C_2H_4 concentration is held constant.

Again, you can use the following expression to solve for any order:

$$\left(\frac{Concentration_{Exp\,A}}{Concentration_{Exp\,B}}\right)^y = \left(\frac{Rate_{Exp\,A}}{Rate_{Exp\,B}}\right)$$

Where y is the order. Thus, ratio the values in experiment 3 to 1.

$$\left(\frac{0.08}{0.02}\right)^y = \left(\frac{0.088}{0.022}\right)$$

$$(4)^y = (4)$$

$$y = 1$$

or

$$\log (2)^x = \log (4)$$

$$y \log 4 = \log 4$$

$$y = \log 4 / \log 4$$

$$y = 1$$

The rate law now goes to:

$$\text{Rate} = k [C_2H_4]^2 [I_2]^1$$

Step 5. Determine the rate constant, k, and write the complete rate law.

To solve for k, one can use any line of experimental data.

Experiment	[C$_2$H$_4$] (M)	[I$_2$] (M)	Rate (M/sec)
1	0.01 M	0.02 M	0.022
2	0.05 M	0.04 M	0.048
3	0.01 M	0.08 M	0.088
4	0.02 M	0.02 M	0.088

Using Experiment 3's data

$$\text{Rate} = k [C_2H_4]^2 [I_2]^1$$

$$0.088 \text{ M/sec} = k [0.01]^2 [0.08]^1$$

$$0.088 \text{ M/sec} = k (8.0 \times 10^{-6} \text{ M}^3)$$

$$k = (0.088 \text{ M/sec}) / (8.0 \times 10^{-6} \text{ M}^3)$$

$$k = 1.1 \times 10^4 \text{ M}^{-2} \text{ sec}^{-1}$$

Thus, the complete rate law for this reaction is:

$$\text{Rate} = 1.1 \times 10^4 \text{ M}^{-2} \text{ sec}^{-1} [C_2H_4]^2 [I_2]^1$$

If ever asked to determine the overall order of the reaction, simply add the orders in the rate law expression. For this reaction, 2 + 1 = 3; thus, the reaction is 3rd order.

STUDENT PROBLEMS

1. The following experimental kinetic data was acquired for the following reaction:

$$2Ti\,(s) + 3\,I_2\,(g) \rightarrow 2\,TiI_3\,(s)$$

Experiment	[Ti] (M)	[I_2] (M)	Rate (M/sec)
1	0.06 M	0.03 M	0.297
2	0.02 M	0.03 M	0.033
3	0.02 M	0.09 M	0.099
4	0.04 M	0.06 M	0.150

What is the rate law for this reaction? What is the overall order of the reaction?

2. The following experimental kinetic data was collected for this reaction:

$$CaCO_3\,(s) + H_2O\,(l) + CO_2\,(g) \rightarrow Ca(HCO_3)_2\,(s)$$

Experiment	[$CaCO_3$] (M)	[H_2O] (M)	[CO_2] (M)	Rate (M/sec)
1	0.01 M	0.02 M	0.01 M	0.003
2	0.02 M	0.01 M	0.01 M	0.012
3	0.01 M	0.01 M	0.01 M	0.006
4	0.01 M	0.01 M	0.02 M	0.024

What is the rate law for this reaction? What is the overall order of the reaction?

(Answers: 1. Rate = 2750 /(M^2sec)[Ti]^2[I_2]^1, overall order = 3;

2. Rate = 60 /(Msec)[$CaCO_3$]^1[H_2O]$^{-1}$[CO_2]^2, overall order = 2)

Section 3.3 Integrated Rate Laws

The integrated set of rate laws incorporate both the concentration of the reactants and time. You can identify these types of kinetic problems by the presence of the time variable in the question. Because 3rd and higher-order reactions rarely occur, this semester of chemistry provides equations for 0, 1st, and 2nd order reactions. The table below shows the integrated rate law and the half-life equations that you will need to solve these types of problems.

Table 3.1 Rate Law Equations

Reaction Order	Rate Law	Integrated Rate Law	Half-Life
0	Rate = k	$[A]_t = -kt + [A]_o$	$\dfrac{[A]_o}{2k}$
1	Rate = k [A]	$\ln\left(\dfrac{[A]_t}{[A]_o}\right) = -kt$	$\dfrac{0.693}{k}$
2	Rate = k [A]^2	$\dfrac{1}{[A]_t} = kt + \dfrac{1}{[A]_o}$	$\dfrac{1}{k[A]_o}$

Solution Tips

1. If the order of the reaction is not stated in the question, you can determine it by either looking at the units of the rate constant or by looking at the rate law (concentration exponent).

2. You cannot set up proportionalities when comparing different points in time, because the amount of reactant always depletes at different concentrations at different times.

3. When you are given percentages, simply convert them to a concentration (e.g., 30% = 30 M).

4. You always need to know the rate constant k.

5. Typically these problems, as presented, will occur in two steps or have two unknowns. You will need to find one variable to use to solve for the other variable.

6. The $[A]_o$ value will always be the greater than $[A]_t$, because the reactant concentration decreases.

Section 3.3.1 First Order Problem

Sample Problem 3.3 Integrated rate laws – first order

The decomposition reaction for SO_2Cl_2 is $SO_2Cl_2 \rightarrow SO_2 + Cl_2$. This reaction is a first-order reaction and is 25% complete in 120 min. How long would it take for the reaction to reach 75% completion?

Step 1. Identify the type of kinetic problem and the known and unknown variables.

In this problem, you have the time and concentration; thus, you will use integrated rate laws.

The decomposition reaction for SO_2Cl_2 is $SO_2Cl_2 \rightarrow SO_2 + Cl_2$. This reaction is a first-order reaction and is 25% complete in 120 min. How long would it take for the reaction to each 75% completion?

The problem states that the reaction is first order, which means you will use this equation:

$$\ln \left(\frac{[A]_t}{[A]_o}\right) = -kt$$

You do not know the time to 75% completion and the rate constant, k. You do know that the reaction is 25% complete in 120 minutes.

Step 2. Solve for k.

Use the 25% complete in 120 mins to solve for the rate constant k. 25% complete means that the reaction started at 100% or 100 M ($[A]_o$) and is now at 75 M (25 M is gone). The 75 M is $[A]_t$.

$$\ln \left(\frac{[A]_t}{[A]_o}\right) = -kt$$

$$\ln \left(\frac{75}{100}\right) = -k\,(120 \text{ minutes})$$

$$-0.2877 = -k\,(120 \text{ min})$$

$$k = -0.2877/-120 \text{ min}$$

$$k = 0.00239 \text{ min}^{-1}$$

Step 3. Use k to solve for time in the second part.

The question asks for the time it takes to reach 75% completion. Again 100 M is $[A]_o$; $[A]_t$ will be 25, because the reaction is 75% complete. The reaction is still first order, so the needed equation is:

$$\ln \left(\frac{[A]_t}{[A]_o}\right) = -kt$$

$$\ln\left(\frac{25}{100}\right) = -(0.00239 \text{ min}^{-1}) \, t$$

$$-1.386 = -(0.00239 \text{ min}^{-1}) \, t$$

$$T = -1.386 / -0.00239 \text{ min}^{-1}$$

$$T = 580 \text{ Minutes}$$

Section 3.3.2 Second Order Problem

Sample Problem 3.4 Integrated rate laws – second order

A certain reaction A →*products* is second order. If this reaction is 65% complete in 30.0 min, what is the half-life?

Step 1. Identify the type of kinetic problem and the known and unknown variables.

In this problem, you have the time and concentration; thus, you will use integrated rate laws.

A certain reaction A →*products* is second order. This reaction reaches 65% completion in 30.0 min, what is the half-life?

You do not know the half-life; you do not know k; and you know that the reaction is 65% complete in 30.0 min. This is also a second order reaction, and you need the following equations to solve the problem:

$$\frac{1}{[A]_t} = kt + \frac{1}{[A]_0} \quad \text{and} \quad t_{1/2} = \frac{1}{k[A]_0}$$

Step 2. Determine the rate constant, k.

Use 65% complete at 30.0 min to solve for the rate constant k. Again, 100% means 100 M, which is $[A]_0$, and $[A]_t$ is 35 M, since the reaction is 65% complete.

$$\frac{1}{[A]_t} = kt + \frac{1}{[A]_0}$$

$$\frac{1}{35M} = k \, (30.0 \text{ minutes}) + \frac{1}{100M}$$

$$0.0286 \text{ M}^{-1} = k \, (30.0 \text{ min}) + 0.01 \text{ M}^{-1}$$

$$0.0286 - 0.01 \text{ M}^{-1} = k \, (30.0 \text{ min})$$

$$k = 0.0186 \text{ M}^{-1} / 30.0 \text{ min}$$

$$k = 0.000619 \text{ M}^{-1} \text{ min}^{-1}$$

Step 3. Using k, solve for the half-life, $t_{1/2}$.

The half-life equation for a second-order reaction is:

$$t_{1/2} = \frac{1}{k[A]_o}$$

$$t_{1/2} = \frac{1}{(0.000619)(100)}$$

$$t_{1/2} = \frac{1}{0.0619}$$

$$t_{1/2} = 16.2 \text{ minutes}$$

STUDENT PROBLEMS

1. A first–order reaction has a rate constant of 0.0033 s^{-1}. What is the time needed for the reaction to go initially from 0.056 M to 0.011 M? What is the half life of this reaction?
2. What is the rate constant for the second-order decomposition reaction when the starting concentration drops by 55.0% after 16.0 minutes? What is the half life?
(Answers: 1. 490 sec, $t_{1/2}$ = 210 sec; 2. 7.60 x 10^{-4} M^{-1}sec^{-1}, 13.2 minutes)

Section 3.4 Temperature and the Effect on Rate: The Arrhenius Equations

The final set of problems presented in this chapter are those that deal with the effect of temperature on the rate of a reaction. You can use the Arrhenius equation to solve these problems:

$$k = Ae^{-E_a/RT}$$

Or you can run a natural log through this equation, which yields:

$$\ln k = \ln A - E_a/RT$$

Where k is the rate constant, A is the frequency factor, R is the ideal gas constant 8.314 J/mol K, T is temperature in Kelvin, and E_a is the activation energy.

This equation is also commonly used in a two-point form, which is very similar to Clausius-Clapeyron Equation from the first chapter about solids and liquids:

$$\ln \frac{k_1}{k_2} = \frac{E_a}{R}\left(\frac{1}{T_1} - \frac{1}{T_2}\right)$$

Solving Tips:

1. To identify the correct form of the Arrhenius equation to use, look to see if the problem mentions the frequency factor A. Only the first form of the equation includes this variable.

2. When using the two-point form of this equation, make sure to keep the correct rate constant with the correct temperature.

3. Make sure to convert the activation energy to Joules in the equation. The Ideal Gas Constant R = 8.314 J/mol K

Sample Problem 3.5 Arrhenius equation – temperature and rate

The activation energy for a particular reaction is 101 kJ/mol. At 25.0°C, the rate constant is 4.0×10^{-3} M^{-1} sec^{-1}; what is the rate constant when the temperature is increased to 68°C?

Step 1. Identify the type of kinetic problem and the known and unknown variables.

The problem asks for activation energy; thus, you must use the Arrhenius equation. Since the problem does not mention the frequency factor and it gives two temperatures, you will need the two-point equation.

$$\ln \frac{k_1}{k_2} = \frac{E_a}{R}\left(\frac{1}{T_1} - \frac{1}{T_2}\right)$$

The activation energy, E_a, for a particular reaction is 101 kJ/mol. At 25.0°C the rate constant is 4.0 x 10^{-3} M^{-1} sec^{-1}; what is the rate constant when the temperature is increased to 68.0°C?

You do not know the second rate constant; the activation energy, the two temperatures, and the first rate constant are known.

Step 2. Convert to the correct units, and plug the known values into the equation.

$$T_1 = 25.0 + 273.15 = 298.15 \text{ K}$$

$$T_2 = 68.0 + 273.15 = 341.15 \text{ K}$$

$$101 \text{ kJ/mol} = 101{,}000 \text{ J/mol}$$

$$k_1 = 4.0 \times 10^{-3} \text{ M}^{-1} \text{ sec}^{-1}$$

Plug the variables into the equation:

$$\ln \frac{k_1}{k_2} = \frac{E_a}{R}\left(\frac{1}{T_1} - \frac{1}{T_2}\right) \text{K}^{-1}$$

$$\ln \frac{4.0 \times 10^{-3} \, M^{-1} s^{-1}}{k_2} = \frac{101000 \, J/mol}{8.314 \, J/molK}\left(\frac{1}{298.15} - \frac{1}{341.15}\right)\text{K}^{-1}$$

Step 3. Simplify the math, and solve for k.

$$\ln \frac{4.0 \times 10^{-3} \, M^{-1} s^{-1}}{k_2} = \frac{101000 \, \cancel{J/mol}}{8.314 \, \cancel{J/molK}}\left(\frac{1}{298.15} - \frac{1}{341.15}\right)\text{K}^{-1}$$

$$\ln \frac{4.0 \times 10^{-3} \, M^{-1} s^{-1}}{k_2} = 12148.18 \text{ K}^{-1} (0.003354 - 0.002931) \text{ K}$$

$$\ln \frac{4.0 \times 10^{-3} \, M^{-1} s^{-1}}{k_2} = 5.1386$$

The next step involves taking the inverse ln of each side:

$$e^{\ln \frac{4.0 \times 10^{-3} \, M^{-1} s^{-1}}{k_2}} = e^{5.1386}$$

$$\cancel{k_2} \frac{4.0 \times 10^{-3} \, M^{-1} s^{-1}}{\cancel{k_2}} = 170.477 \, k_2$$

$$4.0 \times 10^{-3} \text{ M}^{-1} \text{ sec}^{-1} = 170.477 \, k_2$$

$$k_2 = \frac{4.0 \times 10^{-3} M^{-1} sec^{-1}}{170.477}$$

$$k_2 = 2.35 \times 10^{-5} \text{ M}^{-1} \text{ sec}^{-1}$$

Sample Problem 3.6 Arrhenius equation – temperature and rate with frequency factor

The activation energy of a reaction is 200 kJ mol^{-1}, and the frequency factor A is 5.55×10^{19} s^{-1}. What is the value of the rate constant at 500.0°C?

Step 1. Identify the type of kinetic problem and the known and unknown variables.

This problem mentions the frequency factor A; thus, this form of the Arrhenius equation is used:

$$\ln k = \ln A - E_a/RT$$

The activation energy of a reaction is 200 kJ mol^{-1}, and the frequency factor A is 5.55×10^{19} s^{-1}. What is the value of the rate constant at 500.0°C?

The rate constant is unknown; the temperature, frequency factor, and activation energy are known.

Step 2. Convert to the correct units and plug the known values into the equation.

Temperature must be in Kelvin:

$$500.0 + 273.15 = 773.15 \text{ K}$$

Activation energy, E_a, must be in joules.

$$200 \text{ kJ/mol} = 200{,}000 \text{ J/mol}$$

Plug in the known variables:

$$\ln k = \ln A - E_a/RT$$

$$\ln k = \ln(5.55 \times 10^{19} \text{ sec}^{-1}) - \frac{200000 \; J \; mol^{-1}}{(8.314 \; J \; mol^{-1} K^{-1})(773.15 \; K)}$$

$$\ln k = 45.46 - 31.11$$

$$\ln k = 14.35$$

$$k = e^{14.35}$$

$$k = 1.70 \times 10^{6} \text{ sec}^{-1}$$

STUDENT PROBLEMS

1. What is the activation energy for a reaction where the rate constant at 100.0°C is 6.78 x 10^{-4} sec^{-1} and at 350.0°C is 3.20 x 10^{-2} sec^{-1}.
2. The rate constant for a reaction at 400°C is 6.88 × 10^{-6} min^{-1}. What is the frequency factor, A, if the Activation Energy for the reaction is 88.8 kJ/mol?

(Answers: 1. 29.9 kJ/mole; 2. 54)

Section 3.5 Reaction Mechanisms

Reactions often occur in steps. The summation of these steps yields an overall reaction (similar to Hess's law from the first semester). These problems typically ask the student to determine the molecularity (number of reactants), intermediates, catalysts, overall reaction, and the rate determining step.

Sample Problem 3.7 Reaction mechanism

The following steps occur for a particular reaction mechanism.

$$H_2O_2 + I^- \rightarrow HOI + OH^- \quad \text{slow}$$
$$OH^- + H^+ \rightarrow H_2O \quad \text{fast}$$
$$HOI + H^+ + I^- \rightarrow I_2 + H_2O \quad \text{fast}$$

What is the overall reaction? What is the molecularity of each step? What are the intermediates and catalyst for this reaction? What is the rate determining step? What is the rate law for this reaction?

Step 1. Answer each question separately. What is the overall reaction?

To determine the overall reaction, cancel like components to the left and right of the reaction arrow, and then, rewrite the new equation.

$$H_2O_2 + I^- \rightarrow \cancel{HOI} + \cancel{OH^-} \quad \text{slow}$$
$$\cancel{OH^-} + H^+ \rightarrow H_2O \quad \text{fast}$$
$$\cancel{HOI} + H^+ + I^- \rightarrow I_2 + H_2O \quad \text{fast}$$

Overall reaction:

$$H_2O_2 + 2\,I^- + 2\,H^+ \rightarrow 2\,H_2O + I_2$$

Step 2. Determine the molecularity of each elementary reaction.

Remember the overall reaction is not elementary, because it comes from a series of steps. Each step is elementary. So, for this three step reaction:

$H_2O_2 + I^- \rightarrow HOI + OH^-$	Slow – bimolecular – 2 reactants
$OH^- + H^+ \rightarrow H_2O$	Fast – bimolecular – 2 reactants
$HOI + H^+ + I^- \rightarrow I_2 + H_2O$	Fast – termolecular – 3 reactants

Step 3. What are the intermediates and catalyst for this reaction?

Again, the three steps of this reaction must be examined to determine the intermediate and catalyst.

$$H_2O_2 + I^- \rightarrow \cancel{HOI} + \cancel{OH^-} \quad \text{slow}$$
$$\cancel{OH^-} + H^+ \rightarrow H_2O \quad \text{fast}$$
$$\cancel{HOI} + H^+ + I^- \rightarrow I_2 + H_2O \quad \text{fast}$$

When determining the overall reaction, we look to see which components we cancel out to find the intermediates and catalyst. The intermediate is a component that is a product and then a reactant in sequential steps. A catalyst is a canceled component that occurs as a reactant in the first step and then a product in the last step.

In this reaction, HOI and OH^{-1} are intermediates, and a catalyst is not present.

Step 4. What is the rate-determining step? What is the rate law?

The rate determining step for any mechanism is the SLOW step listed in the reaction. You can also use it to define the rate law. For this reaction mechanism:

$$H_2O_2 + I^- \rightarrow \cancel{HOI} + \cancel{OH^-} \quad \text{slow}$$
$$\cancel{OH^-} + H^+ \rightarrow H_2O \quad \text{fast}$$
$$\cancel{HOI} + H^+ + I^- \rightarrow I_2 + H_2O \quad \text{fast}$$

Step 1 is the rate-determining step (slow), and the rate law is:

$$\text{Rate} = k\,[H_2O_2]\,[I^-]$$

STUDENT PROBLEM

1. The following three steps occur for a reaction mechanism:

    Step 1. A + D → B + C (slow)

    Step 2. A + B → C + E (fast)

    Step 3. E → D + I (fast)

What is the overall reaction? What is the molecularity of each step? What are the intermediates and catalyst for this reaction? What is the rate determining step? What is the rate law for this reaction?

2. The following are the two steps for a reaction mechanism:

   Step 1. $2NO_2(g) \rightarrow NO_3(g) + NO(g)$ (slow)
   Step 2. $CO(g) + NO_3(g) \rightarrow CO_2(g) + NO_2(g)$ (fast)

What is the overall reaction? What is the molecularity of each step? What are the intermediates and catalyst for this reaction? What is the rate determining step? What is the rate law for this reaction?

(Answers: 1. 2A → 2C + I, step 1 and 2 are bimolecular, step 3 is unimolecular, Intermediates are B and E, Catalyst is D, Rate determining step is Step 1(slow), Rate = k[A][D]; 2. NO_2 (g) + CO (g) → CO_2 (g) + NO (g), both steps are bimolecular, NO_3 is the intermediate and there is no catalyst, Step 1 is the rate determining step, Rate = $k[NO_2]^2$

Chapter 4. Equilibria

Section 4.1 Determination of K_c and K_p, Equilibrium Concentrations

The first type of equilibrium problem involves the calculation of the equilibrium constant K_c or K_p.

Solving Tips:

1. Only the components that exist in the gaseous and aqueous phase appear in the equilibrium expression.

2. The coefficients preceding each component in the chemical reaction become the exponent in the equilibrium. For the reaction:

$$H_2 \text{ (g)} + I_2 \text{ (g)} \rightleftharpoons 2 HI \text{ (g)}$$

The equilibrium expression would be:

$$K_c = \frac{[HI]^2}{[H_2][I_2]}$$

3. Remember, you solve for the equilibrium constant K_c when you have expressed the components in concentrations and use the K_p when pressures are given. Equilibrium constants do not possess units.

4. Disregard any temperature given in the problem, as you will not use it to solve equilibria.

Sample Problem 4.1 K_c calculations

What is the value of K_c at 25.0°C when the equilibrium concentration of HI is 2.8 M, H_2 is 3.3 M, and I_2 is 5.0 M for the given reaction:

$$H_2 \text{ (g)} + I_2 \text{ (g)} \rightleftharpoons 2 HI \text{ (g)}$$

Step 1. Determine the unknown and known variables.

<u>What is the value of K_c</u> at 25.0°C when the equilibrium concentration of HI is 2.8 M, H_2 is 3.3 M, and I_2 is 5.0 M for the given reaction:

$$H_2 \text{ (g)} + I_2 \text{ (g)} \rightleftharpoons 2 HI \text{ (g)}$$

Step 2. Write the equilibrium expression, and then, fill in the values, and solve for K_c.

$$K_c = \frac{[HI]^2}{[H_2][I_2]}$$

$$K_c = \frac{[2.8\,M]^2}{[3.3M][5.0M]}$$

$$K_c = 0.48$$

Sample Problem 4.2 K_p and equilibrium pressures

The following reaction has a K_p value of 2.8 x 10⁴ at 25.0°C. What is the equilibrium pressure of HI when the pressure for H₂ is 2.60 atm, and I₂ is 1.83 atm?

$$H_2\,(g) + I_2\,(g) \rightleftharpoons 2\,HI\,(g)$$

Step 1. Determine the unknown and known variables.

The following reaction has a K_p value of 2.8 x 10⁴ at 25.0°C. What is the equilibrium pressure of HI when the partial pressure for H₂ is 2.60 atm, and I₂ is 1.83 atm?

$$H_2\,(g) + I_2\,(g) \rightleftharpoons 2\,HI\,(g)$$

You do not know the equilibrium pressure of HI, but you do know the K_p and the equilibrium pressure for H₂ and I₂.

Step 2. Write the equilibrium expression, and then, fill in the values and solve for K_p.

$$K_p = \frac{P_{HI}^2}{P_{H_2}P_{I_2}}$$

$$2.8 \times 10^4 = \frac{P_{HI}^2}{(2.60)(1.83)}$$

$$P_{HI}^2 = (2.8 \times 10^4)(2.60)(1.83)$$

$$P_{HI}^2 = 133224$$

$$\sqrt{P_{HI}^2} = \sqrt{133224}$$

$$P_{HI} = 365\ \text{atm}$$

STUDENT PROBLEMS

1. At 75.0°C, the K_p is 4.6 x 10⁴ for the following reaction:

$$2BrCl_3(g) \rightleftharpoons Br_2(g) + 3Cl_2(g)$$

What is the equilibrium pressure of the BrCl₃ when the equilibrium value for Br₂ is 2.03 atm and for Cl₂ is 4.06 atm?

2. What is the value for K_c for the following reaction at 80.0°C:

$$5A(g) + B(g) \rightleftharpoons 3C(g).$$

When the equilibrium concentration of [A] is 0.55 M, [B] is 0.44 M, and [C] is 0.22 M.
(Answers: 1. 0.054 atm, 2. K_c = 0.48)

Section 4.2 The Relationship between K_c and K_p

The next set of problems mathematically relates K_c and K_p through the equation:

$$K_p = K_c(RT)^{\Delta n}$$

For this equation, R is the gas constant (0.08206 L atm/mol K), and temperature, T, is expressed in Kelvin. Δn is the number of moles of gas in the products minus the number of moles of gas in the reactants. For example, in the reaction:

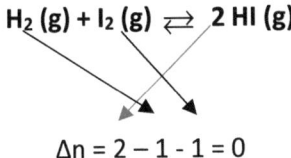

$$\Delta n = 2 - 1 - 1 = 0$$

Sample Problem 4.3 K_p and K_c calculations

K_p for the reaction, $2SO_2(g) + O_2(g) \rightleftharpoons 2SO_3(g)$, is 5×10^{12}. What is K_c for this equilibrium at 40.0°C?

Step 1. Identify the type of equilibrium problem and the known and unknown variables.

This problem gives K_p and asks for K_c; thus, you will use the following equation:

$$K_p = K_c(RT)^{\Delta n}$$

K_p for the reaction, $2SO_2(g) + O_2(g) \rightleftharpoons 2SO_3(g)$, is ($5 \times 10^{12}$) What is K_c for this equilibrium at 40.0°C?

You do not know K_c, but you do know K_p and the temperature.

Step 2. Calculate Δn, fill in the variables, and solve for K_c.

For the given reaction:

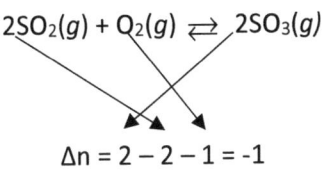

$$\Delta n = 2 - 2 - 1 = -1$$

Solving for K_c:

$$K_p = K_c(RT)^{\Delta n}$$

$$5 \times 10^{12} = K_c \, [(0.08206 \text{ L atm/mol K})(313.15 \text{ K})]^{-1}$$

$$5 \times 10^{12} = K_c (0.0389)$$

$$K_c = 5 \times 10^{12}/0.0389$$

$$K_c = 1.28 \times 10^{14}$$

STUDENT PROBLEMS

1. What is K_c when K_p is 6.50×10^{-4} at 300 K for the reaction, $2NO(g) + Cl_2(g) \rightleftharpoons 2NOCl(g)$?
2. What is K_p at 500°C for the reaction, $3H_2(g) + N_2(g) \rightleftharpoons 2NH_3(g)$, when $K_c = 6.0 \times 10^{-2}$?

(Answers: 1. $K_c = 0.016$; 2. $K_p = 1.7 \times 10^{-5}$)

Section 4.3 Equilibrium Constant Manipulation and the Chemical Equation

Because many reactions occur in steps, you can determine the overall reaction by adding these steps together. The equilibrium constant, K, for this overall reaction can be determined by multiplying the individual K value from each step.

Solving Tips

1. When an equation is flipped (left to right), K becomes K^{-1}.
2. When you use a multiplier 2 throughout the equation, K becomes K^2; If you use 3, the K becomes K^3; and so forth.
3. When you multiply the equation by 1/2, K becomes $K^{1/2}$ or $\sqrt{K}$, if by 1/3 K becomes $K^{1/3}$ or $\sqrt[3]{K}$ and so forth.

Sample Problem 4.4 Equilibrium constant and the chemical equation

K_c for the following reaction at 500°C is 0.06:

$$3H_2(g) + N_2(g) \rightleftharpoons 2NH_3(g)$$

a. What is the value of K_c if the reaction goes to:

$$6H_2(g) + 2N_2(g) \rightleftharpoons 4NH_3(g)$$

b. What is the value of K_c if the reaction goes to:

$$\tfrac{2}{3}NH_3(g) \rightleftharpoons H_2(g) + \tfrac{1}{3}N_2(g)$$

Step 1. Identify the type of equilibrium problem and the known and unknown variables.

When identifying this type of equilibrium problem, always look to see if the chemical equation itself is manipulated mathematically or if it has been flipped (the equation is written in the reverse direction). This problem has two questions, each with a known chemical equation and a known initial value of K. You must determine the new K.

Step 2. Solve for Part A first. Determine the new value of K.

Place the two equations on top of each other so that you can easily see how the original equation compares to the manipulated equation. For these equations, the mathematical change can easily be found.

Original: $3H_2(g) + N_2(g) \rightleftharpoons 2NH_3(g)$

Manipulated: $6H_2(g) + 2N_2(g) \rightleftharpoons 4NH_3(g)$ (multiplied by 2)

By simply looking at one component, you can determine the multiplier. Looking at H_2, the value goes from 3 to 6; thus, the multiplier is 2.

If the multiplier is 2, then K goes to K^2.

Original K = 0.06

New K = Original $K^2 = (0.06)^2 = 0.0036$

Step 3. Solve for Part B. Determine the new value of K.

Utilize the same reasoning as in Step 2. Compare the original to the manipulated equation:

Original: $3H_2(g) + N_2(g) \rightleftharpoons 2NH_3(g)$

Manipulated: $\frac{2}{3}NH_3(g) \rightleftharpoons H_2(g) + \frac{1}{3}N_2(g)$

The first thing you will notice when comparing the original to the manipulated equation is the H_2 was a reactant in the first, and it became a product in the manipulated reaction. When this occurs, the equation has been flipped, and the original K goes to K^{-1}.

Next, the coefficients do not match from equation to equation; therefore, you need to determine the multiplier. Again, you only need to check one component. If you select the H_2 again, the coefficient goes from 3 to 1; thus, the original equation was DIVIDED by 3.

The original K, which in the first manipulation became K^{-1}, now becomes $K^{-1/3}$ (multiply the mathematical steps, or $K^{-1} \times K^{1/3} = K^{-1/3}$)

Original K = 0.06

New K = Original $K^{-1/3} = (0.06)^{-1/3} = 2.55$

STUDENT PROBLEMS

1. The equilibrium constant for the reaction $N_2(g) + O_2(g) \rightleftharpoons 2NO(g)$ is 8.80×10^{-4}.
   What is K for the reaction:

   $3N_2(g) + 3O_2(g) \rightleftharpoons 6NO(g)$

   What is K for the reaction:

   $NO(g) \rightleftharpoons \frac{1}{2}N_2(g) + \frac{1}{2}O_2(g)$

2. Using the following set of equations and their equilibrium constants, what is the value of K for the overall reaction:

$$4A + B \rightleftharpoons 2E$$

Reaction Step 1. $4A + D \rightarrow C$
Reaction Step 2. $2E + G \rightarrow B$
Reaction Step 3. $3G + 3C \rightarrow 3D$

(Answers: 1a. K = 6.81 x 10^{-10}, 1b. K = 33.7; 2. $K_{overall} = (K_{Step\ 1})(1/K_{Step2})(\sqrt[3]{K_{Step\ 3}})$

Section 4.4 The Use of the Reaction Quotient, Q

In this set of equilibrium problems, you do not know the equilibrium status of the chemical reaction at the concentration of the given components. To identify these problems, look for the wording "Is this reaction at equilibrium? If not, in which direction must it be shifted?"

Solving Tips

1. To identify these problems, look for the wording "Is this reaction at equilibrium? If not, in which direction must it be shifted?"
2. The written equilibrium expression is the same for Q and K. K is at equilibrium; Q is out of equilibrium.
3. Compare the numerical values of Q and K:

If Q < K, then the reaction must shift to the right to get into equilibrium

If Q > K, then the reaction must shift to the left to get into equilibrium

If Q = K, equilibrium is established

Sample Problem 4.5 Reaction quotient calculations

The K_p is 80.6 at 600°C for this reaction:

$$H_2(g) + I_2(g) \rightleftharpoons 2HI(g)$$

Is the system at equilibrium when 0.0187 atm of hydrogen, 0.122 atm of iodine, and 2.08 atm of hydrogen iodide are present? If not, which way must the reaction shift to reach equilibrium?

Step 1. Identify the type of equilibrium problem and the known and unknown variables.

The K_p is 80.6 at 600°C for this reaction:

$$H_2(g) + I_2(g) \rightleftharpoons 2HI(g)$$

Is the system at equilibrium when 0.0187 atm of hydrogen, 0.122 atm of iodine, and 2.08 atm of hydrogen iodide are present? If not, which way must the reaction shift to reach equilibrium?

This problem asks if the reaction is at equilibrium; thus, Q will be assessed versus the given K. You know K, you know the concentration of each component, and you do not know the equilibrium status.

Step 2. Write the equilibrium expression, and solve for Q.

$$K_p = \frac{P_{HI}^2}{P_{H_2}P_{I_2}} \text{ or } Q = \frac{P_{HI}^2}{P_{H_2}P_{I_2}}$$

$$Q = \frac{(2.08)^2}{(0.0187)(0.122)}$$

$$Q = 1896$$

Step 3. Compare Q versus K, determine the equilibrium status, and predict the direction the equation will shift in order to reach equilibrium

$$Q = 1896$$

$$K = 80.6$$

$$1896 > 80.6$$

Chemical reaction is not at equilibrium, because $Q \neq K$.

$$Q > K$$

Thus, the reaction must shift to the left to reestablish equilibrium.

STUDENT PROBLEMS

1. For the reaction $H_2(g) + Br_2(g) \rightleftharpoons 2HBr(g)$, $K_c = 156$ at 500°C. The concentration of $[H_2]$ is 0.0033 M, $[Br_2]$ is 0.0088 M, and $[HBr]$ is 0.0055 M. Is the system at equilibrium? If not, which direction must the reaction shift to reach equilibrium?

2. K_c for the following equation at 100°C is 1.02×10^{-3}:

$$A(g) + 3B(g) \rightleftharpoons 4C(g).$$

   Is this reaction at equilibrium if [A] = 1.8 M, [B] = 1.3 M, and [C] = 0.25 M? If not, in which direction must the reaction shift to reach equilibrium?

   (Answers: 1. Q = 1.04, 1.04 < 156 Equilibrium shifts Right; 2. Q = 9.88×10^{-4}, $9.88 \times 10^{-4} < 1.02 \times 10^{-3}$ equilibrium shifts right)

Section 4.5 Finding Equilibrium Concentrations, ICE Tables

The final type of equilibrium problem typically asks for the determination of the equilibrium concentration for a reactant or product. You can readily identify these types of problems, because you are given the initial amounts and are typically asked to solve for the equilibrium amount. These problems have numerous question styles, which make these types the biggest challenge in this chapter.

Finally, these type of Initial, Change, Equilibrium (ICE) problems will be used in future chapters for acid/base and solubility equilibria. Thus, it is very important to have a strong knowledge of how to solve the problems, since they will appear again!

Solving Tips

1. Make sure all reactants and products are expressed in concentration (M) or pressure.
2. Make sure to assess which way the reaction will shift when it proceeds.
3. Make sure to include the coefficient in the reaction as part of the change condition. These are stoichiometric relationships.
4. In these problems, you are always solving for the change or value of x. Thus, make sure to see how the equilibrium value for each component is expressed in terms of x.
5. If a polynomial results, you must use the quadratic equation to solve for the value of x:

$$\text{Quadratic Equation: } ax^2 + bx + c = 0$$

Solution:

$$X = \frac{-b \pm \sqrt{b^2 - 4ac}}{2a}$$

6. If the problem asks for a partial pressure, that is the equilibrium condition for the component expressed in terms of the change, x.
7. The total pressure of the system is the sum of all partial pressures of all gases at equilibrium
8. If all components of a given reaction have initial concentrations, you may not readily see the shift to equilibrium. When this scenario occurs, you must calculate Q and assess this value versus the given K. Remember, if Q = K, equilibrium exists; if Q < K, the reaction shifts to the right, and if Q > K, the reaction shifts left.

Sample Problem 4.6 Calculation of equilibrium concentrations and ICE tables

For the reaction:

$$2HI(g) \rightleftarrows H_2(g) + I_2(g)$$

$K_c = 0.0288$ at 300°C. Initially, 2.3 M of HI was placed into the reaction vessel. What is the equilibrium concentration of H_2?

Step 1. Identify the type of equilibrium problem and the known and unknown variables.

Because the problem asks for an equilibrium concentration, and an initial state is given, this will be an ICE table problem.

For the reaction:

$$2HI(g) \rightleftarrows H_2(g) + I_2(g)$$

$K_c = 0.0288$ at 300°C. Initially, 2.3 M of HI was placed into the reaction vessel. What is the equilibrium concentration of H_2?

You know the equation, K_c value, and initial concentration of HI; you do not know the concentration of H_2 at equilibrium.

Step 2. Write the equilibrium expression for the reaction, and create an ICE table.

$$K_c = \frac{[H_2][I_2]}{[HI]^2}$$

$$2HI(g) \rightleftarrows H_2(g) + I_2(g)$$

Initial

Change

Equilibrium

Step 3. Fill in the initial conditions, the change, and the equilibrium values

Initially, there is only 2.3 M of HI. The H₂ and I₂ have not formed yet.

	2HI(g) ⇌	H₂(g) +	I₂(g)
Initial	2.3 M	0	0
Change			
Equilibrium			

The equilibrium will shift in the direction where a zero exists. Thus, this reaction will shift right.

	2HI(g) ⇌	H₂(g) +	I₂(g)
Initial	2.3 M	0	0
	SHIFT →		
Change			
Equilibrium			

Assess the Change, x

Because the equilibrium shifts right, HI will deplete, and the other two components will form. Thus, the change in HI is negative, and the change in H₂ and I₂ are positive. Remember to include the coefficient prior to each component in the change (e.g., 2HI means a loss of 2x).

	2HI(g) ⇌	H₂(g) +	I₂(g)
Initial	2.3 M	0	0
	SHIFT →		
Change	-2x	+x	+x
Equilibrium			

Write the equilibrium condition.

	2HI(g) ⇌	H₂(g) +	I₂(g)
Initial	2.3 M	0	0
	SHIFT →		
Change	-2x	+x	+x
Equilibrium	2.3 – 2x	x	x

Once you have completed the table, you can now solve the problem for x.

Step 4. Place the equilibrium values from the ICE table into the equilibrium expression, and solve for x.

	2HI(g) ⇌	H₂(g) +	I₂(g)
Initial	2.3 M	0	0
	SHIFT →		
Change	-2x	+x	+x
Equilibrium	2.3 – 2x	x	x

$$K_c = \frac{[H_2][I_2]}{[HI]^2}$$

$$0.0288 = \frac{(x)(x)}{(2.3-2x)^2}$$

$$0.0288 = \frac{x^2}{(2.3-2x)^2}$$

Upon initial inspection of this mathematical expression, a polynomial exists. However, you can simplify the right-hand side of the equation, because the expressions in both the numerator and denominator are squared. Thus, to simplify, you take the square root of both sides, simplifying the expression and eliminating the polynomial.

$$\sqrt{0.0288} = \sqrt{\frac{x^2}{(2.3-2x)^2}}$$

$$0.1697 = \frac{x}{2.3-2x}$$

$$(2.3-2x)\,0.1697 = \frac{x}{\cancel{2.3-2x}}\,\cancel{(2.3-2x)}$$

$$(2.3-2x)\,0.1697 = x$$

On the left side of the expression, the 0.1697 is distributed and this expression results:

$$0.3903 - 0.3394x = x$$

$$0.3903 - 0.3394x + 0.3394x = x + 0.3394x$$

$$0.3903 = 1.3394\,x$$

$$x = \frac{0.3903}{1.3394} = 0.291$$

Step 5. Finally, determine the equilibrium concentration of H_2.

Inspection of the ICE table shows that, at equilibrium, the concentration of H_2 equals x. Thus, $[H_2]$ at equilibrium is 0.291M.

Sample Problem 4.7 Calculation of equilibrium concentrations and ICE tables

(Initial values for all substances are given)

For the reaction:

$$2HI(g) \rightleftarrows H_2(g) + I_2(g)$$

K_c = 0.0288 at 300ºC. Initially, 2.3 M of HI, 2.0 M of H_2, and 2.0 M of I_2 were placed into the reaction vessel. What is the equilibrium concentration of HI?

Step 1. Identify the type of equilibrium problem and the known and unknown variables.

Because the problem asks for an equilibrium concentration, and an initial state is given, this will be an ICE table problem.

For the reaction:

$$2HI(g) \rightleftarrows H_2(g) + I_2(g)$$

$K_c = 0.0288$ at 300°C. Initially, 2.3 M of HI, 2.0 M of H_2 and 2.0 M of I_2 were placed into the reaction vessel. What is the equilibrium concentration of HI?

You know the equation, K_c value, and initial concentration of all three components; you do not know the concentration of HI at equilibrium.

Step 2. Write the equilibrium expression for the reaction, and create an ICE table.

$$2HI(g) \rightleftharpoons H_2(g) + I_2(g)$$

Initial

Change

Equilibrium

$$K_c = \frac{[H_2][I_2]}{[HI]^2}$$

Step 3. Fill in the initial conditions, the change, and the equilibrium values.

Initially, there is 2.3 M of HI, 2.0M of H_2, and 2.0 M of I_2.

$$2HI(g) \rightleftharpoons H_2(g) + I_2(g)$$

Initial	2.3 M	2.0 M	2.0 M

Change

Equilibrium

In this problem, all three components have initial concentrations. Therefore, the shift is not readily seen. When this scenario occurs, the reaction quotient, Q, must be calculated and assessed versus the given K. Remember, if Q = K, equilibrium exists; if Q < K, the reaction shifts to the right, and if Q > K, the reaction shifts left.

K for this reaction is 0.0288.

$$Q = \frac{[H_2][I_2]}{[HI]^2}$$

$$Q = \frac{(2)(2)}{[2.3]^2}$$

$$Q = 0.756$$

$$K = 0.0288$$

Thus, Q > K

Reaction shifts to the left

	2HI(g)	⇌	H₂(g)	+ I₂(g)
Initial	2.3 M		2.0 M	2.0 M
	SHIFT	←		
Change				
Equilibrium				

Assess the change, x

Because the equilibrium shifts left, HI forms, and the other two components deplete. Thus, the change in HI is positive, and the change in H_2 and I_2 are negative. Remember to include the coefficient prior to each component in the change (e.g., 2HI means a gain of 2x).

	2HI(g)	⇌	H₂(g)	+ I₂(g)
Initial	2.3 M		2.0 M	2.0 M
	SHIFT	←		
Change	+2x		-x	-x
Equilibrium				

Write the equilibrium condition

	2HI(g)	⇌	H₂(g)	+ I₂(g)
Initial	2.3 M		2.0 M	2.0 M
	SHIFT	←		
Change	+2x		-x	-x
Equilibrium	2.3 + 2x		2.0-x	2.0-x

Once you have completed the table, you can now solve the problem for x.

Chapter 4. Equilibria | 73

Step 4. Place the equilibrium values from the ICE table into the equilibrium expression, and solve for x.

	2HI(g) ⇌	H$_2$(g)	+ I$_2$(g)
Initial	2.3 M	2.0	2.0
	SHIFT ←		
Change	+2x	-x	-x
Equilibrium	2.3 + 2x	2.0-x	2.0-x

$$K_c = \frac{[H_2][I_2]}{[HI]^2}$$

$$0.0288 = \frac{(2-x)(2-x)}{(2.3+2x)^2}$$

$$0.0288 = \frac{(2-x)^2}{(2.3+2x)^2}$$

Upon initial inspection of this mathematical expression, a polynomial exists. However, you can simplify the right-hand side of the equation, because the expressions in both the numerator and denominator are squared. Thus, to simplify, take the square root of both sides, simplifying the expression and eliminating the polynomial.

$$\sqrt{0.0288} = \sqrt{\frac{(2-x)^2}{(2.3+2x)^2}}$$

$$0.1697 = \frac{2-x}{2.3+2x}$$

$$(2.3 + 2x)\, 0.1697 = \frac{2-x}{2.3+2x}\, \cancel{(2.3+2x)}$$

$$(2.3 + 2x)\, 0.1697 = 2 - x$$

On the left side of the expression, the 0.1697 is distributed and this expression results:

$$0.3903 + 0.3394x = 2 - x$$

$$0.3903 + 0.3394x - 0.3394x = 2 - x - 0.3394x$$

$$0.3903 = 2 - 1.3394\,x$$

$$-1.6097 = -1.3394\,x$$

$$X = \frac{-1.6097}{-1.3394} = 1.202$$

Step 5. Finally, determine the equilibrium concentration of HI.

Inspection of the ICE table shows that, at equilibrium, the concentration of HI equals 2.3M + 2x. With x = 1.202, the [HI] at equilibrium is:

$$[HI] = 2.3\,M + 2(1.202) = 4.7\,M$$

Sample Problem 4.8. Calculation of equilibrium pressures and ICE tables

$$2HI(g) \rightleftarrows H_2(g) + I_2(g)$$

For this reaction, K_p = 0.0333 at 600°C. 1.50 atm of HI was injected into a container, and the system was allowed to reach equilibrium. What is the equilibrium partial pressure of H_2? What is the total pressure of the system at equilibrium?

Step 1. Identify the type of equilibrium problem and the known and unknown variables.

Because the problem asks for an equilibrium partial pressure and gives an initial state, this will be an ICE table problem.

For the reaction:

$$2HI(g) \rightleftarrows H_2(g) + I_2(g)$$

For this reaction, K_p = 0.0333 at 600°C. 1.50 atm of HI was injected into a container, and the system was allowed to reach equilibrium. What is the equilibrium partial pressure of H_2? What is the total pressure of the system at equilibrium?

You know the equation, K_p, and the initial pressure of HI; you do not know the equilibrium partial pressure of H_2 and the total pressure at equilibrium.

Step 2. Write the equilibrium expression for the reaction, and create an ICE table.

$$K_p = \frac{P_{H_2} P_{I_2}}{P_{HI}^2}$$

Chapter 4. Equilibria | 75

$$2HI(g) \rightleftharpoons H_2(g) + I_2(g)$$

Initial

Change

Equilibrium

Step 3. Fill in the initial conditions and the change and the equilibrium values.

Initially, there is 1.50 atm of HI.

	$2HI(g)$	$\rightleftharpoons$	$H_2(g)$	$+ I_2(g)$
Initial	1.50 atm		0	0
Change				
Equilibrium				

The equilibrium will shift in the direction where a zero exists in the initial condition of the ICE table. Thus, this reaction will shift right.

	$2HI(g)$	$\rightleftharpoons$	$H_2(g)$	$+ I_2(g)$
Initial	1.50 atm		0	0
	SHIFT	→		
Change				
Equilibrium				

Assess the change, x.

Because the equilibrium shifts right, HI depletes, and the other two components form. Thus, the change in HI is negative, and the change in H_2 and I_2 are positive. Remember to include the coefficient prior to each component in the change (e.g., 2HI means a loss of 2x).

	2HI(g)	⇌	H₂(g)	+ I₂(g)
Initial	1.5 atm		0	0
	SHIFT	→		
Change	-2x		+x	+x
Equilibrium				

Write the equilibrium condition.

	2HI(g)	⇌	H₂(g)	+ I₂(g)
Initial	1.5 atm		0	0
	SHIFT	→		
Change	-2x		+x	+x
Equilibrium	1.5 - 2x		+x	+x

Once you complete the table, you can now solve the problem for x.

Step 4. Place the equilibrium values from the ICE table into the equilibrium expression, and solve for x.

	2HI(g)	⇌	H₂(g)	+ I₂(g)
Initial	1.5 atm		0	0
	SHIFT	→		
Change	-2x		+x	+x
Equilibrium	1.5 - 2x		+x	+x

$$K_p = \frac{P_{H_2} P_{I_2}}{P_{HI}^2}$$

$$0.0333 = \frac{(x)(x)}{(1.5-2x)^2}$$

$$0.0333 = \frac{x^2}{(1.5-2x)^2}$$

Upon initial inspection of this mathematical expression, a polynomial exists. However, it can be simplified on the right hand side of the equation, because the expressions in both the numerator and denominator are squared. Thus, to simplify, take the square root of both sides, simplifying the expression and eliminating the polynomial.

$$\sqrt{0.0333} = \sqrt{\frac{x^2}{(1.5-2x)^2}}$$

$$0.1825 = \frac{x}{1.5-2x}$$

$$(1.5 + 2x)\,0.1825 = \frac{x}{\cancel{1.5-2x}}\cancel{(1.5-2x)}$$

$$(1.5 - 2x)\,0.1825 = x$$

On the left side of the expression, the 0.1825 is distributed which results in the expression:

$$0.2737 - 0.3650x = x$$

$$0..2737 + 0.3650x - 0.3650x = x + 0.3650x$$

$$0.2737 = 1.3650\,x$$

$$X = \frac{0..2737}{1.3650} = 0.196 \text{ atm}$$

Step 5. Finally, determine the partial pressure of H_2 and the total pressure.

Inspection of the ICE table shows that, at equilibrium, the concentration of H_2 equals x. With x = 0.2914, the equilibrium partial pressure is:

$$P_{H_2} = x = 0.196 \text{ atm}$$

The total pressure is the sum of all the partial pressures at equilibrium of each component.

Thus,

$$P_{H_2} = x = 0.196 \text{ atm}$$

$$P_{I_2} = x = 0.196 \text{ atm}$$

$$P_{HI} = 1.5 - 2x = 1.5 - 2(0.2196) = 1.11 \text{ atm}$$

$$P_{tot} = P_{H_2} + P_{I_2} + P_{HI}$$

$$P_{tot} = 0.196 + 0.196 + 1.11 = 1.50 \text{ atm}$$

Sample Problem 4.9 Calculation of equilibrium pressures and ICE tables – quadratic and non-equilibria components

For the reaction:

$$CO_2 (g) + C (s) \rightleftharpoons 2 CO (g)$$

The K_p = 1.17 at 1000°C. What is the equilibrium partial pressure of CO when 5.0 moles of C and 2.66 atm of CO are initially introduced into a 5.0 L reaction vessel?

Step 1. Identify the type of equilibrium problem and the known and unknown variables.

Because the problem asks for an equilibrium partial pressure and an initial state is given, this will be an ICE table problem.

For the reaction:

$$CO_2 (g) + C (s) \rightleftharpoons 2 CO (g)$$

The K_p = 1.17 at 1000°C. What is the equilibrium partial pressure of CO when 5.0 moles of C and 2.66 atm of CO are initially introduced into a 5.0 L reaction vessel?

You know he equation, K_p, and the initial pressure of CO_2; you do not know the equilibrium partial pressure of CO. The value of C is not necessary, because it is a solid, which is not included in equilibrium expression.

Step 2. Write the equilibrium expression for the reaction, and create an ICE table.

$$K_p = \frac{P_{CO}^2}{P_{CO_2}}$$

	$CO_2 (g)$	+ C (s)	$\rightleftharpoons$	2 CO (g)
Initial				
Change				
Equilibrium				

Chapter 4. Equilibria | 79

Step 3. Fill in the initial conditions and the change and the equilibrium values.

Initially, there is 2.66 atm of CO. The 5 moles of C are excluded because C is a solid.

	CO_2 (g)	+ C (s)	⇌	2 CO (g)
Initial	0			2.66 atm
Change				
Equilibrium				

Not included in ICE table or equilibrium

The equilibrium will shift in the direction where a zero exists in the initial condition of the ICE table. Thus, this reaction will shift left.

	CO_2 (g)	+ C (s)	⇌	2 CO (g)
Initial	0			2.66 atm
			← SHIFT	
Change				
Equilibrium				

Assess the change, x.

Because the equilibrium shifts left, CO depletes, and CO_2 forms. Thus, the change in CO is negative, and the change in CO_2 is positive. Remember to include the coefficient prior to each component in the change (e.g., 2 CO means a loss of 2x).

	CO_2 (g)	+ C (s)	⇌	2 CO (g)
Initial	0			2.66 atm
			← SHIFT	
Change	+x			-2x
Equilibrium				

Write the equilibrium condition.

	CO_2 (g)	+ C (s)	⇌	2 CO (g)
Initial	0			2.66 atm
			←	SHIFT
Change	+x			-2x
Equilibrium	x			2.66 − 2x

Once you complete the table, you can now solve the problem for x.

Step 4. Place the equilibrium values from the ICE table into the equilibrium expression, and solve for x.

	CO_2 (g)	+ C (s)	⇌	2 CO (g)
Initial	0			2.66 atm
			←	SHIFT
Change	+x			-2x
Equilibrium	x			2.66 − 2x

$$K_p = \frac{P_{CO}^2}{P_{CO_2}}$$

$$1.17 = \frac{(2.66 - 2x)^2}{(x)}$$

Upon initial inspection of this mathematical expression, a polynomial exists. However, you cannot simplify this equation. Thus, you use the quadratic equation to find x.

Quadratic Equation: $ax^2 + bx + c = 0$

Solution:

$$x = \frac{-b \pm \sqrt{b^2 - 4ac}}{2a}$$

Solving for x and expressing the quadratic:

$$1.17x = (2.66 - 2x)(2.66 - 2x)$$

Distribute the parentheses on the right:

$$1.17x = 7.08 - 5.32x - 5.32x + 4x^2$$

$$1.17x - 1.17x = 7.08 - 10.64x + 4x^2 - 1.17x$$

$$0 = 7.08 - 11.81x + 4x^2$$

Thus a = 4, b = -11.81, and c = 7.08 for the quadratic solution.

$$x = \frac{-b \pm \sqrt{b^2 - 4ac}}{2a}$$

Fill in the values for a, b, and c, and solve for x:

$$x = \frac{11.81 \pm \sqrt{-11.81^2 - 4(4)(7.08)}}{2(4)}$$

$$x = \frac{11.81 \pm \sqrt{139.5 - 113.3}}{8}$$

$$x = \frac{11.81 \pm \sqrt{26.2}}{8}$$

$$x = \frac{11.81 + 5.12}{8} \text{ and } x = \frac{11.81 - 5.12}{8}$$

$$x = \frac{11.81 + 5.12}{8} \text{ and } x = \frac{11.81 - 5.12}{8}$$

$$x = 2.12 \text{ and } 0.836$$

Step 5. Finally, determine the partial pressure of CO.

Inspection of the ICE table shows that, at equilibrium, the concentration of CO equals:

$$P_{CO} = 2.66 - 2x$$

With x = 2.12 and 0.836, solve for the equilibrium partial pressure using both values:

$$P_{CO} = 2.66 - 2x$$

$$P_{CO} = 2.66 - 2(2.12) = -1.58$$

or

$$P_{CO} = 2.66 - 2(0.836) = 0.988$$

Note: Pressures and concentrations can never be a negative value. Thus, the correct answer is:

$$P_{CO} = 2.66 - 2(0.836) = 0.988$$

Sample Problem 4.10 Calculation of equilibrium concentrations and ICE tables – determining an initial value

$K_c = 1.37 \times 10^{-4}$ at 1065°C for the reaction:

$$2H_2S(g) \rightleftarrows 2H_2(g) + S_2(g)$$

Initially, an unknown amount of H_2S is placed into a container and allowed to come to equilibrium. At equilibrium, the $[H_2] = 0.044M$; what is the initial concentration of H_2S?

Step 1. Identify the type of equilibrium problem and the known and unknown variables.

This type of problem poses a slight twist as to the way equilibria problems are commonly presented. In this case, you know the equilibrium concentration and will solve for the initial concentration. Even though you do not know the initial concentrations, the problem still includes equilibrium and initial conditions and, therefore, is an ICE table problem.

$K_c = 1.37 \times 10^{-4}$ at 1065°C for the reaction:

$$2H_2S(g) \rightleftarrows 2H_2(g) + S_2(g)$$

Initially, an unknown amount of H_2S is placed into a container and allowed to come to equilibrium. At equilibrium, the $[H_2] = 0.044M$; what is the initial concentration of H_2S?

You know the equation, K_c, value and equilibrium concentration of H_2; you do not know the initial concentration of H_2S.

Step 2. Write the equilibrium expression for the reaction, and create an ICE table.

$$K_c = \frac{[S_2][H_2]^2}{[H_2S]^2}$$

Chapter 4. Equilibria | 83

$$2H_2S(g) \rightleftharpoons 2H_2(g) + S_2(g)$$

Initial

Change

Equilibrium

Step 3. Fill in the known conditions and the change and the equilibrium values.

Since x represents a change, use y as the unknown variable for the initial concentration of H_2S. The initial values for the other two components are 0, since initially, only H_2S is introduced into the container

	$2H_2S(g)$	$\rightleftharpoons$	$2H_2(g)$	$+ S_2(g)$
Initial	y		0	0
Change				
Equilibrium				

The equilibrium will shift in the direction where a zero exists in the initial condition of the ICE table. Thus, this reaction will shift right.

	$2H_2S(g)$	$\rightleftharpoons$	$2H_2(g)$	$+ S_2(g)$
Initial	y		0	0
	SHIFT	→		
Change				
Equilibrium				

Assess the change, x.

Because the equilibrium shifts right, H_2S will deplete, and the other two components will form. Thus, the change in H_2S is negative, and the changes in H_2 and S_2 are positive. Remember to include the coefficient prior to each component in the change. E.g., 2 H_2S means a loss of 2x.

	$2H_2S(g)$	⇌	$2H_2(g)$	$+ S_2(g)$
Initial	y		0	0
	SHIFT	→		
Change	-2x		+2x	+x
Equilibrium				

Write the equilibrium condition.

	$2H_2S(g)$	⇌	$2H_2(g)$	$+ S_2(g)$
Initial	y		0	0
	SHIFT	→		
Change	-2x		+2x	+x
Equilibrium	y − 2x		+2x	+x

Here's the key to the problem, the value of x is known when an equilibrium concentration is provided in the problem. At equilibrium, $[H_2] = 0.044M$, which means:

	$2H_2S(g)$	⇌	$2H_2(g)$	$+ S_2(g)$
Initial	y		0	0
	SHIFT	→		
Change	-2x		+2x	+x
Equilibrium	y − 2x		+2x = 0.044M	+x

Solving for x and expressing components at equilibrium:

$$+2x = 0.044 M$$

$$x = 0.022 \text{ M}$$

Thus, at equilibrium:

$$[H_2] = 0.044 M$$

$$[S_2] = 0.022 \text{ M}$$

$$[H_2S] = y - 2(0.022M) = y - 0.044M$$

Step 4. Rewrite the equilibrium expression, and solve for Y, the initial concentration.

$$K_c = \frac{[S_2][H_2]^2}{[H_2S]^2}$$

$$1.37 \times 10^{-4} = \frac{(0.022)(0.044)^2}{(Y-0.044)^2}$$

$$1.37 \times 10^{-4} (y-0.044)^2 = 4.26 \times 10^{-5}$$

$$(y-0.044)^2 = 0.311$$

$$(y-0.044)(y-0.044) = 0.311$$

$$y^2 - 0.044y - 0.044y + 0.0019 = 0.311$$

$$y^2 - 0.088y - 0.3091 = 0$$

Thus, a = 1, b = -0.088, and c = -0.3091 for the quadratic solution.

$$y = \frac{-b \pm \sqrt{b^2 - 4ac}}{2a}$$

$$y = \frac{0.088 \pm \sqrt{0.0077 + 1.2364}}{2}$$

$$y = \frac{0.088 \pm 1.1154}{2}$$

$$y = \frac{0.088 + 1.1154}{2} \text{ and } y = \frac{0.088 - 1.1154}{2}$$

$$y = 0.602 \text{ and } -0.5134$$

Since a concentration cannot be negative, the initial concentration of H_2S is 0.602M.

STUDENT PROBLEMS

1. $K_c = 100$ at 600 K for the following equilibrium: $X(g) + Y(g) \rightleftharpoons 2Z(g)$. Determine the equilibrium concentration of Z if 2 moles of X and 2 moles of Y are introduced to a 500.0 mL container?

2. $K_p = 0.011$ at 1000°C for the reaction $N_2(g) + O_2(g) \rightleftharpoons 2NO(g)$. Initially, 4.0 atm of N_2 and 4.0 atm of O_2 are placed into a 2.0 L container. What is the partial pressure of NO at equilibrium? What is the total pressure of the system when it reaches equilibrium?

3. For the reaction of the decomposition of sodium bicarbonate at 250°C:

$$2NaHCO_3(s) \rightleftharpoons Na_2CO_3(s) + CO_2(g) + H_2O(g)$$

   K_p equals 0.50. What is the partial pressure of CO_2 at equilibrium, if 10.0 grams of $NaHCO_3$ is initially placed a reaction vessel flask? What will the total gas pressure at equilibrium?

4. $K_c = 55.5$ for the reaction:

$$O_2(g) + H_2(g) \rightleftharpoons CO(g) + H_2O(g)$$

   Initially, if 4.00 mol of oxygen and 4.00 mol of hydrogen are placed in a 1.0 L container, what will be the concentration of CO at equilibrium?

   (Answers: 1. 6.67 M; 2. P_{NO} = 0.398 atm, P_{total} = 8.0 atm; 3. P_{CO2} = 0.71 atm, P_{total} = 1.42 atm; 4. 3.93 M)

Section 4.6 Le Chatelier's Principle

Le Chatelier's principle is a conceptual method of predicting which way a reaction in chemical equilibrium will shift when changes in concentration, temperature, and pressure occur.

Solving Tips

1. Adding a component shifts the reaction in the opposite direction.
2. Decreasing the amount of a component shifts the reaction to the side of the chemical equation that component is located.
3. Increasing pressure decreases volume. Thus, the equilibrium shifts to the side with a smaller number of gas moles.
4. Decreasing pressure increases volume. Thus, the equilibrium shifts to the side with more gas moles.
5. When looking at temperature changes, the enthalpy of the reaction will be given in the question. Remember a negative enthalpy means an exothermic reaction. Thus, heat goes on the product side. If enthalpy is positive, it is an endothermic reaction, and heat goes onto the reactant side of the reaction.
6. Once the enthalpy is placed on the correct side, an increase in temperature shifts the equilibrium away from that side. A decrease in temperature shifts the reaction to the side on which the component is located.
7. Remember a solid or liquid state substance is not part of an equilibrium expression, and changes to their concentration will have no effect or shift to the equilibrium.
8. Catalysts speed up a chemical reaction and will affect the forward and reverse reactions equally. Thus, addition of a catalyst will have no effect on an equilibrium.

Sample Problem 4.11 Applying Le Chatelier's principle to chemical equilibria

For the reaction:

$$3H_2(g) + N_2(g) \rightleftharpoons 2NH_3(g) \quad \Delta H = -45.9 \text{ kJ/mol}$$

Answer the following:

a. Which way does the equilibrium shift if the temperature is increased?
b. Which way does the equilibrium shift if the pressure is increased?
c. Which way does the equilibrium shift if the $[H_2]$ is decreased?

Step 1. Identify the type of equilibrium problem and the known and unknown variables.

Since temperature, pressure, and a concentration is being affected and the equilibrium direction shift is asked for, this problem is a Le Chatelier's Principle problem; you know the equation and the enthalpy of the reaction, and the unknowns are summarized in steps a, b, and c.

Step 2. Determine the reaction side for the enthalpy and solve questions.

The enthalpy is negative; thus, the reaction is exothermic. Heat goes to the right, or product, side.

$$3H_2(g) + N_2(g) \rightleftharpoons 2NH_3(g) + 45.9 \text{ kJ/mol}$$

a. Which way does the equilibrium shift if the temperature increases?

Temperature increases, so the reaction will shift to the left to reestablish the equilibrium.

$$3H_2(g) + N_2(g) \rightleftharpoons 2NH_3(g) + \mathbf{45.9 \text{ kJ/mol}\uparrow}$$

← **SHIFT LEFT**

b. Which way does the equilibrium shift if the pressure increases?

A pressure increase means volume decreases. Thus, the shift is towards the side with a smaller number of gas moles.

$$3H_2(g) + N_2(g) \rightleftharpoons 2NH_3(g) + 45.9 \text{ kJ/mol}$$
4 gas moles　　　2 gas moles

SHIFT RIGHT →

c. Which way does the equilibrium shift if the $[H_2]$ decreases?

$$\downarrow 3H_2(g) + N_2(g) \rightleftharpoons 2NH_3(g) + 45.9 \text{ kJ/mol}$$

← **SHIFT LEFT**

STUDENT PROBLEMS

1. Answer the following questions in regards to this chemical reaction:

$$C\,(s) + CO_2\,(g) \rightleftharpoons CO\,(g); \quad \Delta H = 283 \text{ kJ/mol}$$

   a. Which way does the equilibrium shift if the temperature is increased?
   b. Which way does the equilibrium shift if the pressure is decreased?
   c. Which way does the equilibrium shift if the [C] is decreased?
   d. Which way does the equilibrium shift if the [CO] is decreased?
   e. Which way does the equilibrium shift if a catalyst were added?
   (Answers: a. right; b. no change; c. no change; d. right; e. no change)

Chapter 5. Acids and Bases

This Chapter presents a variety of different problems dealing with Bronsted-Lowery acids and bases. By definition, a Bronsted-Lowery acid DONATES a proton, while a Bronsted-Lowery base ACCEPTS the proton in these reactions. The best way to ensure a successful solution of any of these problems is to be able to identify if the problem has a strong or weak acid or base or is a salt in solution.

Section 5.1 K_w Equilibrium, pH and pOH Problems

The auto-ionization equilibrium of water provides the basic equations utilized to determine the concentrations of hydronium ion, H_3O^+, and hydroxide ion, OH^-. These values then lead to the determination of pH and pOH of acidic and basic solutions. The following set of equations summarizes what you need to solve many of these problems.

$$H_2O + H_2O \rightleftharpoons H_3O^+ + OH^-$$

At 25°C, $K_w = 1.0 \times 10^{-14} = [OH^-] \times [H_3O^+]$

$$pH = -\log[H_3O^+]$$

$$pOH = -\log[OH^-]$$

$$pH + pOH = 14$$

$$[H_3O^+] = 10^{-pH}$$

$$[OH^-] = 10^{-pOH}$$

Sample Problem 5.1 $[OH^-]$ and $[H_3O^+]$

What is the $[OH^-]$ for a solution for which $[H_3O^+] = 1.44 \times 10^{-8}$?

Step 1. Identify the equation and known and unknown variables.

Since $[OH^-]$ and $[H_3O^+]$ are presented in the problem, use the following equation:

$$K_w = 1.0 \times 10^{-14} = [OH^-] \times [H_3O^+]$$

You know K_w and $[H_3O^+]$, and you do not know $[OH^-]$.

Step 2. Fill in the known variables and solve for $[OH^-]$.

$$1.0 \times 10^{-14} = [OH^-] \times [1.44 \times 10^{-8}]$$

$$1.0 \times 10^{-14} = [OH^-] \times [1.44 \times 10^{-8}]$$

$$[OH^-] = \frac{1.0 \times 10^{-14}}{1.44 \times 10^{-8}}$$

$$[OH^-] = 6.94 \times 10^{-7}$$

Sample Problem 5.2. Concentration to pH and pOH

What is the pH and pOH of the solution in sample problem 5.1?

$$[H_3O^+] = 1.44 \times 10^{-8} \text{ and } [OH^-] = 6.94 \times 10^{-7}$$

Step 1. Identify the needed equations and the known and unknown variables.

You must determine the pH and pOH; thus, the needed equations are:

$$pH = -\log[H_3O^+]$$

$$pOH = -\log[OH^-]$$

You do not know the pH and pOH, but you do know the $[H_3O^+]$ and $[OH^-]$.

Step 2. Fill in the known variables, and solve.

You can take two different paths to achieve the results for these types of problems.

Path 1. Use pH and pOH equations.

$$pH = -\log[H_3O^+]$$

$$pH = -\log(1.44 \times 10^{-8})$$

$$pH = 7.84$$

$$pOH = -\log[OH^-]$$

$$pOH = -\log(6.94 \times 10^{-7})$$

$$pOH = 6.16$$

Path 2. Solve for either the pH or pOH, and then use pH + pOH = 14.

$$pH = -\log[H_3O^+]$$

$$pH = -\log(1.44 \times 10^{-8})$$

$$pH = 7.84$$

$$pH + pOH = 14$$

$$7.84 + pOH = 14$$

$$pOH = 14 - 7.84$$

$$pOH = 6.16$$

Sample Problem 5.3. pH and pOH to concentration problem

What is the $[H_3O^+]$ for a solution with a pH = 4.55?

Step 1. Identify the needed equations and the known and unknown variables.

The question asks you to determine the $[H_3O^+]$ from the pH; thus, you need the following equation:

$$[H_3O^+] = 10^{-pH}$$

Step 2. Fill in the known variables, and solve.

$$[H_3O^+] = 10^{-pH}$$

$$[H_3O^+] = 10^{-4.55}$$

$$[H_3O^+] = 2.82 \times 10^{-5}$$

STUDENT PROBLEMS

1. What is the $[H_3O^+]$ and pH of a solution with a $[OH^-] = 3.33 \times 10^{-4}$?
2. What is the $[OH^-]$ of a solution with a pH = 10.99?
3. What is the pOH of a solution with a $[H_3O^+] = 6.34 \times 10^{-3}$?
   (Answers: 1. 3.00×10^{-11}, pH = 10.52; 2. 9.77×10^{-4}; 3. 11.80)

Section 5.2 Strong Acids and Strong Bases

Problems involving strong acid and strong bases utilize the set of equations listed in the previous section:

$$pH = -\log[H_3O^+]$$

$$pOH = -\log[OH^-]$$

$$pH + pOH = 14$$

$$[H_3O^+] = 10^{-pH}$$

$$[OH^-] = 10^{-pOH}$$

Students are urged to memorize both lists of strong acids and strong bases. Any other acids and or bases encountered in problems will then be weak.

Strong Acids:

HNO_3, H_2SO_4, $HClO_4$, $HClO_3$, HCl, HBr, HI

Strong Bases:

All Group 1A Hydroxides: LiOH, NaOH, KOH, RbOH, and CsOH,

Heavy Group 2A Hydroxides: $Ca(OH)_2$, $Sr(OH)_2$ and $Ba(OH)_2$

Sample Problem 5.4 pH determination of strong acids

What is the pH of 0.0020 M solution of HNO_3?

Step 1. Identify the needed equation for solution, and the known and unknown variables.

Solve for the pH from the concentration of HNO_3—a strong acid; thus, you need the following equation: $pH = -\log[H_3O^+]$

Note: A STRONG ACID concentration is equal to $[H_3O^+]$. Thus, in this problem, the 0.0020 M HNO_3 possesses $[H_3O^+] = 0.0020$ M. You do not know the pH.

Step 2. Fill in the known variables, and solve.

$$pH = -\log[H_3O^+]$$

$$pH = -\log[0.0020]$$

$$pH = 2.70$$

Sample Problem 5.5 pH determination of strong bases

What is the pH of a 0.00030 M solution of $Ca(OH)_2$?

Step 1. Identify the needed equation for solution, and the known and unknown variables.

Solve for the pH from the concentration of $Ca(OH)_2$—a strong base; thus, you will need the following equation:

$$pOH = -\log[OH^-]$$

Note: A STRONG BASE concentration is equal to $[OH^-]$. Thus, in this problem, the 0.0003 M $Ca(OH)_2$ possesses $[OH^-]$= 0.0006M, because in the dissociation of $Ca(OH)_2$, 2 OH^- form in the reaction is:

$$Ca(OH)_2 \rightarrow Ca^{+2} + 2OH^-$$

Thus, by stoichiometry, the OH^- amount will be 2 x .00030 M $Ca(OH)_2$ = 0.00060 M.

You do not know the pH.

Step 2. Fill in the known variables, and solve.

$$pOH = -\log[OH^-]$$

$$pOH = -\log[0.00060]$$

$$pOH = 3.22$$

$$pOH + pH = 14$$

$$pH + 3.22 = 14$$

$$pH = 14 - 3.22 = 10.78$$

Note: The Group 1A Strong Bases dissociate in a 1:1 ratio; thus, a 0.010 M LiOH solution contains 0.010 M $[OH^-]$.

STUDENT PROBLEMS

1. What is the pH of a 0.22 M solution of KOH?
2. What is the pOH of a 2.22 x10^{-4} M solution of HI?
   (Answers: 1. 13.34; 2. 10.35)

Section 5.3 Weak Acids and Bases

Weak acids and bases do not 100% disassociate in water. You must remember your strong acids and bases; all others are weak and will utilize these solution methods. Thus, you cannot readily determine the pH of these solutions by the pH and pOH equations just utilized to solve for strong acids and bases. The following acids and bases form equilibria in aqueous solutions:

Weak Acid:

$$HNO_2\ (aq) + H_2O\ (l) \rightleftarrows H_3O^+\ (aq) + NO_2^{-1}\ (aq)$$

and Weak Base:

$$NH_3\ (aq) + H_2O\ (l) \rightleftarrows OH^-\ (aq) + NH_4^{+1}\ (aq)$$

Since equilibria now occur, you can write an equilibrium expression for each reaction, and two new equilibrium constants now exist. For the two equilibria above:

Weak Acid:

$$K_a = \frac{[H_3O^+][NO_2^{-1}]}{[HNO_2]}$$

Weak Base:

$$K_b = \frac{[NH_4^{+1}][OH^{-1}]}{[NH_3]}$$

As seen in these equilibria, when the change, x, is solved for, either the [H_3O^+] or the [OH^-] will be determined. Thus, you can readily calculate the pH or the pOH.

Solving Tips

1. Weak Acids can be both inorganic and organic compounds. Inorganic Acids will follow the Arrhenius definition of an acid, for which the chemical formula begins with a H. For example, HF, H_3PO_4, H_2SO_3.
2. Organic weak acids are commonly carboxylic acids with the chemical formula XCOOH. Where X can be any carbon containing group. For example, acetic acid is CH_3COOH (or $C_2H_4O_2$), formic acid HCOOH ($HCHO_2$), or benzoic acid C_6H_5COOH.
3. Weak Bases can also be inorganic and organic, as well. Inorganic bases will also follow the Arrhenius definition and possess an OH group at the end of the chemical formula, for example, $Mg(OH)_2$.
4. Organic weak bases are commonly amines with the chemical formula XN, where X can be any carbon containing group, and the N is found at the end of the chemical formula, for example, aniline $C_6H_5NH_2$, pyridine C_5H_5N, and urea NH_2CONH_2.

5. Di and polyprotic acids require multiple ICE tables to solve for pH. Each Hydrogen must be considered seperately

Sample Problem 5.6 pH of a Weak Acid

What is the pH of a 0.18 M solution of HNO_2? K_a for $HNO_2 = 4.5 \times 10^{-4}$.

Step 1. Identify the type of acid or base problem and the known and unknown variables.

Because the problem asks for the pH and a K_a is given, this is a weak acid equilibrium and an ICE table problem.

What is the pH of a 0.18 M solution of HNO_2? K_a for $HNO_2 = 4.5 \times 10^{-4}$?

You know the equation, K_a, value and initial concentration of HNO_2; you do not know the pH. Remember to get the pH you need to determine the H_3O^+ concentration.

Step 2. Write the equilibrium and equilibrium expression and create an ICE table.

This is a weak acid equilibrium, where the HNO_2 is a Bronsted acid, which donates a proton in this reaction:

$$HNO_2\ (aq) + H_2O\ (l) \rightleftharpoons H_3O^+\ (aq) + NO_2^{-1}(aq)$$

$$K_a = \frac{[H_3O^+][NO_2^{-1}]}{[HNO_2]}$$

$$HNO_2\ (aq) + H_2O\ (l) \rightleftharpoons H_3O^+(aq) + NO_2^{-1}(aq)$$

Initial

Change

Equilibrium

Step 3. Fill in the initial conditions, the change, and the equilibrium values.

Initially, there is only 0.18 M of HNO_2. H_2O is in a liquid sate and is excluded from the equilibrium expression.

	HNO₂ (aq)	+ H₂O (l)	⇌	H₃O⁺(aq) +	NO₂⁻¹(aq)
Initial	0.18 M			0	0
Change					
Equilibrium					

The equilibrium will shift in the direction where a zero exists. Thus, this reaction will shift right.

	HNO₂ (aq)	+ H₂O (l)	⇌	H₃O⁺(aq)+	NO₂⁻¹(aq)
Initial	0.18 M			0	0
	SHIFT		→		
Change					
Equilibrium					

Assess the change, x.

Because the equilibrium shifts right, HNO₂ will deplete, and the other two components will form. Thus, the change in HNO₂ is negative, and the change in H₃O⁺ and NO₂⁻¹ are positive.

	HNO₂ (aq)	+ H₂O (l)	⇌	H₃O⁺(aq) +	NO₂⁻¹(aq)
Initial	0.18 M			0	0
	SHIFT		→		
Change	-x			+x	+x
Equilibrium					

Write the equilibrium condition.

	HNO_2 (aq)	+ H_2O (l)	⇌	H_3O^+(aq) +	NO_2^{-1}(aq)
Initial	0.18 M			0	0
Change	-x			+x	+x
Equilibrium	0.18 - x			+x	+x

Once you have completed the table, you can now solve the problem for x.

Step 4. Place the equilibrium values from the ICE table into the equilibrium expression, and solve for x.

$$K_a = \frac{[H_3O^+][NO_2^{-1}]}{[HNO_2]}$$

$$4.5 \times 10^{-4} = \frac{(x)(x)}{(0.18-x)}$$

$$4.5 \times 10^{-4} = \frac{(x)^2}{(0.18-x)}$$

Upon first inspection of this equation, a polynomial exists. You can absolutely solve these problems using the quadratic equation. However, the K_a's for many of the weak acids are very small numbers. Thus, the value of x will often be very small, as well. Actually, if small enough, you can deem it negligible and exclude it from the equilibrium expression.

Many textbooks indicate that, to assess if it is negligible, calculate the percent ionization at the end of solving the problem. Here, we suggest an alternative: assessing the negligibility after writing the equilibrium expression.

Step 5. Assess x negligibility.

Use the following equation to determine the negligibility of x:

$$\frac{Initial\ Concentration\ of\ Weak\ Acid}{K_a\ of\ weak\ acid} > 100$$

For this problem:

$$\frac{0.18\ M}{4.5 \times 10^{-4}} > 100$$

$$400 > 100$$

Here, 400 is greater than 100; thus, x is negligible and can be excluded from the equilibrium expression.

Step 6. Solve for x, and calculate pH.

$$4.5 \times 10^{-4} = \frac{(x)^2}{(0.18-x)}$$

$$4.5 \times 10^{-4} = \frac{(x)^2}{(0.18)}$$

$$8.1 \times 10^{-5} = x^2$$

$$\sqrt{8.1 \times 10^{-5}} = \sqrt{x^2}$$

$$9.0 \times 10^{-3} = x$$

In this problem, x = [H$_3$O$^+$] (weak acid hydrolysis); thus, the pH is:

$$pH = -\log[H_3O^+]$$

$$pH = -\log[9.0 \times 10^{-3}]$$

$$pH = 2.05$$

Percent Disassociation

As mentioned above, you also determine percent disassociation in these sets of questions. Use the following equation to solve for this quantity:

$$\% \text{ Disassociation} = \frac{[H_3O^+] \text{ at equilibrium}}{[HNO_2] \text{ initial}} \times 100\%$$

$$\% \text{ Disassociation} = \frac{9.0 \times 10^{-3}}{0.18} \times 100\%$$

$$\% \text{ Disassociation} = 5.0\%$$

Sample Problem 5.7 pH determination of a weak base

What is the pH of a 0.033 M solution of NH$_3$? K$_b$ for NH$_3$ = 1.8 x 10^{-5}.

Step 1. Identify the type acid or base problem and the known and unknown variables.

Because the problem asks for the pH and gives a K$_b$, this is a weak base equilibrium and an ICE table problem.

What is the pH of a 0.033 M solution of NH$_3$? K$_b$ for NH$_3$ = 1.8 x 10^{-5}.

You know the equation, K_b value, and initial concentration of NH_3; you do not know the pH.

Step 2. Write the equilibrium and equilibrium expression, and create an ICE table.

This is a weak base equilibrium, where the NH_3 is a Bronsted base, which accepts a proton in this reaction:

$$NH_3\ (aq) + H_2O\ (l) \rightleftarrows OH^-\ (aq) + NH_4^{+1}(aq)$$

$$K_b = \frac{[NH_4^{+1}][OH^{-1}]}{[NH_3]}$$

	NH_3 (aq)	+ H_2O (l)	$\rightleftarrows$	OH^- (aq)	+ NH_4^{+1}(aq)
Initial					
Change					
Equilibrium					

Step 3. Fill in the initial conditions, the change, and the equilibrium values.

Initially, there is only 0.033 M of NH_3. H_2O is in a liquid state and is excluded from the equilibrium expression.

	NH_3 (aq)	+ H_2O (l)	$\rightleftarrows$	OH^- (aq)	+ NH_4^{+1}(aq)
Initial	0.033 M			0	0
Change					
Equilibrium					

The equilibrium will shift in the direction where a zero exists. Thus, this reaction will shift right.

	NH₃ (aq)	+ H₂O (l)	⇌	OH⁻ (aq)	+ NH₄⁺¹ (aq)
Initial	0.033 M			0	0
SHIFT			→		
Change					
Equilibrium					

Assess the change, x.

Because the equilibrium shifts right, NH₃ will deplete, and the other two components will form. Thus, the change in NH₃ is negative, and the change in NH₄⁺¹ and OH⁻¹ are positive.

	NH₃ (aq)	+ H₂O (l)	⇌	OH⁻ (aq)	+ NH₄⁺¹ (aq)
Initial	0.033 M			0	0
SHIFT			→		
Change	-x			+x	+x
Equilibrium					

Write the equilibrium condition.

	NH₃ (aq)	+ H₂O (l)	⇌	OH⁻ (aq)	+ NH₄⁺¹ (aq)
Initial	0.033 M			0	0
SHIFT			→		
Change	-x			+x	+x
Equilibrium	0.033 - x			+x	+x

Once you have completed the table, you can now solve the problem for x.

Step 4. Place the equilibrium values from the ICE table into the equilibrium expression, and solve for x.

$$K_b = \frac{[NH_4^{+1}][OH^{-1}]}{[NH_3]}$$

$$1.8 \times 10^{-5} = \frac{(x)(x)}{(0.033-x)}$$

$$1.8 \times 10^{-5} = \frac{(x)^2}{(0.033-x)}$$

Upon first inspection of this equation, a polynomial exists. You can absolutely solve these problems using the quadratic equation. However, the K_b's for many of the weak bases are very small numbers. Thus, the value of x will often be very small, as well. Actually, if small enough, you can deem it negligible and exclude it from the equilibrium expression.

Many textbooks indicate that, in order to assess if it is negligible, calculate the percent ionization at the end of solving the problem. Here, we suggest an alternative, which includes assessing negligibility after writing the equilibrium expression.

Step 5. Assess x negligibility.

Use the following method to see if you can deem x negligible:

$$\frac{Initial\ Concentration\ of\ Weak\ Base}{K_b\ of\ weak\ base} > 100$$

For this problem:

$$\frac{0.033\ M}{1.8 \times 10^{-5}} > 100$$

$$1833 > 100$$

Here, 1833 is greater than 100; thus, x is negligible, and you can exclude it from the equilibrium expression.

Step 6. Solve for x, and calculate pOH.

For weak base equilibria, the x in the expression stands for the [OH⁻]. Thus, you can determine pOH and then the pH:

$$1.8 \times 10^{-5} = \frac{(x)^2}{(0.033-\cancel{x})}$$

$$1.8 \times 10^{-5} = \frac{(x)^2}{(0.033)}$$

$$5.94 \times 10^{-7} = x^2$$

$$\sqrt{5.94 \times 10^{-7}} = \sqrt{x^2}$$

$$7.71 \times 10^{-4} = x$$

In this problem, x = [OH⁻] (weak base hydrolysis); thus, the pOH is:

$$pOH = -\log[OH^-]$$

$$pOH = -\log[7.71 \times 10^{-4}]$$

$$pOH = 3.11$$

Remember:

$$pH + pOH = 14$$

$$pH + 3.11 = 14$$

$$pH = 10.89$$

Sample Problem 5.8 Starting with pH or x, calculate K_a or K_b

What is the K_a of a 0.010 M solution of a weak acid with a pH = 3.33?

Step 1. Identify the type of acid or base problem and the known and unknown variables.

Because the problem asks for the K_a, and you have the pH; this is a weak acid equilibrium and an ICE table problem.

What is the K_a of a ⟨0.010 M solution of a weak acid (HA) with a pH = 3.33?⟩

You know the equation, pH, and initial concentration of a weak acid; you do not know the K_a. Remember to get the pH you need to determine the H_3O^+ concentration, and if you have the pH, you can solve for the H_3O^+ concentration.

Step 2. Write the equilibrium and equilibrium expression, and create an ICE table.

This is a weak acid equilibrium, and a Bronsted acid donates a proton in this reaction:

$$HA\ (aq) + H_2O\ (l) \rightleftharpoons H_3O^+\ (aq) + A^{-1}(aq)$$

$$K_a = \frac{[H_3O^+][A^{-1}]}{[HA]}$$

$$HA\,(aq) + H_2O\,(l) \rightleftharpoons H_3O^+(aq) + A^{-1}(aq)$$

Initial

Change

Equilibrium

Step 3. Fill in the initial conditions, the change, and the equilibrium values.

Initially, there is only 0.010 M of HA. H$_2$O is in a liquid state and is excluded from the equilibrium expression.

	HA (aq)	+ H$_2$O (l) ⇌	H$_3$O$^+$(aq) +	A^{-1}(aq)
Initial	0.010 M		0	0
	SHIFT	→		
Change				
Equilibrium				

The equilibrium will shift in the direction where a zero exists. Thus, this reaction will shift right.

Assess the change, x.

Because the equilibrium shifts right, HA will deplete, and the other two components will form. Thus, the change in HA is negative, and the change in H$_3$O$^+$ and A^{-1} are positive.

	HA (aq)	+ H$_2$O (l) ⇌	H$_3$O$^+$(aq) +	A^{-1}(aq)
Initial	0.010 M		0	0
	SHIFT	→		
Change	-x		+x	+x
Equilibrium				

Write the equilibrium condition.

	HA (aq)	+ H₂O (l)	⇌	H₃O⁺(aq) +	A⁻¹(aq)
Initial	0.010 M			0	0
	SHIFT		→		
Change	-x			+x	+x
Equilibrium	0.010 - x			+x	+x

Step 4. Define x and solve for K_a.

You know the pH (3.33); thus, you also know the H_3O^+ concentration at equilibrium by:

$$[H_3O^+] = 10^{-pH}$$

$$[H_3O^+] = 10^{-3.33}$$

$$[H_3O^+] = 4.68 \times 10^{-4}$$

Thus:
$$[H_3O^+] = x = 4.68 \times 10^{-4}$$

Now you can solve for K_a since x is known:

$$K_a = \frac{[H_3O^+][A^{-1}]}{[HA]}$$

$$K_a = \frac{(x)(x)}{(0.01-x)}$$

$$K_a = \frac{x^2}{(0.01-x)}$$

$$K_a = \frac{(4.68 \times 10^{-4})^2}{(0.01-4.68 \times 10^{-4})}$$

$$K_a = \frac{2.19 \times 10^{-7}}{0.00953}$$

$$K_a = 2.30 \times 10^{-5}$$

STUDENT PROBLEMS

1. What is the K_b of a 0.22 M solution of a weak base with a pH = 10.55?
2. What is the pH of a 0.044 M solution of Aniline ($C_6H_5NH_2$)? K_b for Aniline = 4.2×10^{-10}?
3. What is the pH of a 0.34 M solution of HClO? What is the percent disassociation? K_a of HClO = 3.5×10^{-8}.
(Answers: 1. 5.72×10^{-7}; 2. 8.63; 3. pH = 3.96, % disassociation = 0.032%)

Section 5.4 pH of Salts in Solution

Section 5.4.1 K_a, K_b, and K_w

Weak acids or bases produce strong conjugates, and these conjugates will hydrolyze. When looking at these two reactions, the summation results in the auto-ionization of water, which is K_w. For example:

Step 1.   ~~HF (aq)~~ + H_2O (l) ⇌ ~~F⁻ (aq)~~ + H_3O^+ (aq) K_a

Step 2.   ~~F⁻ (aq)~~ + H_2O (l) ⇌ ~~HF (aq)~~ + OH^- (aq) K_b

Summation: $2\ H_2O$ (l) ⇌ OH^- (aq) + H_3O^+ (aq) K_w

Thus: $K_a \times K_b = K_w$

This equation becomes integral in the determination of pH for salts in solution.

One point of emphasis, a weak acid and its conjugate base, or a weak base and its conjugate acid, are forever linked in all of these hydrolysis reactions. Basically, if the problem has one, it forms the other with the addition of water.

For example, reacting the acetate ion $C_2H_3O_2^{-1}$ with H_2O forms acetic acid $C_2H_4O_2$. Reacting acetic acid with water creates the acetate ion. Reacting NH_3 with water creates the ammonium ion NH_4^{+1}, while reacting NH_4^{+1} with water will produce NH_3.

Sample Problem 5.9 pH determination of a conjugate ion

What is the K_b for CN^{-1}? K_a for HCN = 4.9×10^{-10}.

Step 1. Identify the type of acid/base problem, and determine known and unknown variables.

This problem asks for K_b and gives K_a; thus, the equation $K_a \times K_b = K_w$ will be utilized. Remember, $K_w = 1.0 \times 10^{-14}$.

<u>What is the K_b for CN^-</u>? K_a for HCN = 4.9×10^{-10}.

Step 2. Plug in the known variables, and solve for K_b:

$$K_a \times K_b = K_w$$

$$4.9 \times 10^{-10} \times K_b = 1.0 \times 10^{-14}$$

$$K_b = \frac{1.0 \times 10^{-14}}{4.9 \times 10^{-10}}$$

$$K_b = 2.04 \times 10^{-5}$$

STUDENT PROBLEM

1. What is the K_a for $C_2H_5NH_3^{+1}$? K_b for $C_2H_5NH_2 = 4.7 \times 10^{-4}$. (Answer: 2.13×10^{-11})

Section 5.4.2 pH of Salt Solutions

The key in recognizing these types of problems is to be able to identify that the question has a salt in the problem. A salt is a cation and anion in combination or, simply, a metal and a non-metal. When salts are placed into water, they can form either neutral, acidic, or basic solutions. Identification of the strength can be determined by adding an OH group to the cation and a H before the anion.

Section 5.4.2.1 Salts from Strong/Strong Acid/Base Systems

Strong acids and bases form weak conjugates, which DO NOT hydrolyze. Therefore, these salts that form from the strong acid and bases will not affect pH.

For example:

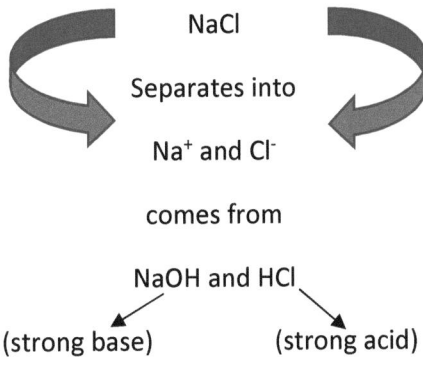

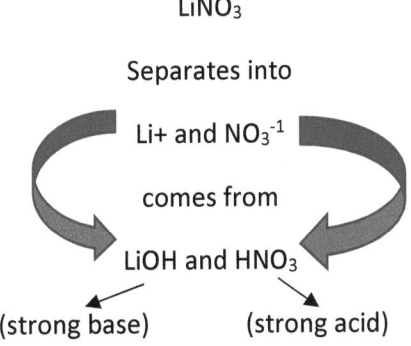

Na^+, Cl^-, Li^+, and NO_3^{-1} all come from strong acids and bases, which make each ion a weak conjugate that does not hydrolyze. Therefore, these salts form neutral solutions.

Section 5.4.2.2 Salts from Weak and Strong Acid/Base Systems

The pH of these salt solutions, will be either acidic or basic depending on which part of the salt (cation or anion) is from a weak acid or base.

Basic Solutions:

For Example:

$$NaNO_2$$

Separates into

$$Na^+ \text{ and } NO_2^-$$

comes from

$$NaOH \text{ and } HNO_2$$

Strong base and weak acid

NO_2^- is the strong conjugate base of the weak acid HNO_2; therefore, the NO_2^- will hydrolyze and affect the pH. Conjugate bases form basic solutions. The Na^+ comes from a strong base and is a weak conjugate. Thus, this cation has no effect on the pH.

Therefore, $NaNO_2$ forms a basic solution.

Acidic Solutions:

$$NH_4Cl$$

Separates into

$$NH_4^+ \text{ and } Cl^-$$

Which comes from

$$NH_4OH \text{ and } HCl$$

weak base and strong acid

NH_4^+ is the strong conjugate acid of the weak acid NH_3, and it will hydrolyze to form an acidic solution when dissolved in water. The Cl^- comes from the strong acid HCl and is a weak conjugate, which has no effect on pH.

In summary, cations that are strong conjugates form acidic solutions (amines), and anions that are strong conjugates form basic solutions.

Section 5.4.2.3 Salts from Weak and Weak Acid/Base Systems

The pH of these systems is determined by looking at the K_a and K_b values for both the cation and anion. Whichever is greater will dictate the pH of the resulting solution.

For Example:

$$NH_4CN$$

Separates into

$$NH_4^+ \text{ and } CN^-$$

Which comes from

$$NH_4OH \text{ and } HCN$$

Weak base and weak acid

The K_a for the strong conjugate acid NH_4^+ = 5.6 x 10^{-10}

$$K_a = \frac{K_w}{K_b} = \frac{1.0 \times 10^{-14}}{1.8 \times 10^{-5}} = 5.6 \times 10^{-10}$$

The K_b for the strong conjugate base CN^- = 2.0 x 10^{-5}

$$K_b = \frac{K_w}{K_a} = \frac{1.0 \times 10^{-14}}{4.9 \times 10^{-10}} = 2.0 \times 10^{-5}$$

Since the K_b (2.0 x 10^{-5}) value is greater than the K_a (5.6 x 10^{-10}) value, this solution will be basic.

Section 5.4.2.4 pH Determination of a Salt Solution

The previous three sections define the method by which you can determine the acidity or basicity of a solution. This section will define the methodology used to calculate the pH of a salt solution. These problems will again utilize an ICE table similar to those used in the beginning sections of this chapter.

Sample Problem 5.10 pH of a salt solution

What is the pH of a 0.010 M solution of NaCN. The K_a for HCN is 4.9×10^{-10}.

Step 1. Identify the type of acid/base problem and the known and unknown variables.

What is the pH of a 0.010 M solution of NaCN. The K_a for HCN is 4.9×10^{-10}

NaCN is a salt (Na^+ and CN^-); thus, this is a salt solution acid/base problem. You know the concentration of the salt and the K_a of the weak acid HCN. You will determine the pH.

Step 2. Write the equilibrium for the strong conjugate, and create an ICE table.

For this problem, Na^+ does not affect pH, it is only affected by the strong conjugate base, CN^-. Thus, after calculating the pH, it should be basic.

$$CN^- (aq) + H_2O (l) \rightleftharpoons HCN (aq) + OH^- (aq)$$

$$K_b = \frac{[HCN][OH^{-1}]}{[CN^{-1}]}$$

	CN^- (aq)	+ H_2O (l)	↔	OH^- (aq) +	HCN (aq)
Initial					
Change					
Equilibrium					

Step 2. Complete the ICE table for the shift, the change, x, and the equilibrium conditions.

	CN^- (aq)	+ H_2O (l)	⇌	OH^- (aq) +	HCN (aq)
Initial	0.010 M			0	0
Change					
Equilibrium					

This equilibrium shifts to the right, because the products were not present initially.

	CN⁻ (aq)	+ H₂O (l) ⇌	OH⁻ (aq) +	HCN (aq)
Initial	0.010 M		0	0
	SHIFTS		→	
Change	-x		+x	+x
Equilibrium	0.010 - x		x	x

Once you have completed the table, you can now solve the problem for x.

Step 3. Place the equilibrium values from the ICE table into the equilibrium expression, and solve for x.

$$K_b = \frac{[HCN][OH^{-1}]}{[CN^{-1}]}$$

At this point, you do not know the K_b, but you know K_a. Recall $K_a \times K_b = K_w$, and you must utilize this to prove the K_b value for the CN⁻ ion.

$$\frac{K_w}{K_a} = K_b = \frac{[HCN][OH^{-1}]}{[CN^{-1}]}$$

$$\frac{K_w}{K_a} = K_b = \frac{[HCN][OH^{-1}]}{[CN^{-1}]}$$

$$K_b = \frac{1.0 \times 10^{-14}}{4.9 \times 10^{-10}} = \frac{[HCN][OH^{-1}]}{[CN^{-1}]}$$

$$K_b = \frac{1.0 \times 10^{-14}}{4.9 \times 10^{-10}} = \frac{[HCN][OH^{-1}]}{[CN^{-1}]}$$

$$2.0 \times 10^{-5} = \frac{[x][x]}{[0.01-x]}$$

Upon first inspection of this equation, a polynomial exists, and you can absolutely use the quadratic equation to solve for x. However, you can check the value of x to see if it is negligible by the method performed in section 4.3.

Step 4. Assessment of x negligibility

Use the following equation to see if you can deem x negligible:

$$\frac{Initial\ Concentration\ of\ Weak\ Base}{K_b\ of\ weak\ base} > 100$$

For this problem:

$$\frac{0.010\ M}{2.0 \times 10^{-10}} > 100$$

$$5.0 \times 10^7 > 100$$

As shown, 5.0×10^7 is greater than 100; thus, x is negligible and can be excluded from the equilibrium expression.

Step 5. Solve for x, and calculate pOH.

For weak base equilibria, the x in the expression stands for the [OH⁻]. Thus, you can determine the pOH then the pH.

$$2.0 \times 10^{-5} = \frac{(x)^2}{(0.01-\cancel{x})}$$

$$2.0 \times 10^{-5} = \frac{(x)^2}{(0.01)}$$

$$2.0 \times 10^{-7} = x^2$$

$$\sqrt{2.0 \times 10^{-7}} = \sqrt{x^2}$$

$$4.5 \times 10^{-4} = x$$

In this problem, x = [OH⁻] (strong conjugate base hydrolysis); thus, the pOH is:

$$pOH = -\log[OH^-]$$

$$pOH = -\log[4.5 \times 10^{-4}]$$

$$pOH = 3.35$$

Remember:

$$pH + pOH = 14$$

$$pH + 3.35 = 14$$

$$pH = 10.65$$

STUDENT PROBLEMS

1. What is the pH of a 0.088 M solution of NH_4Cl? K_b for $NH_3 = 1.8 \times 10^{-5}$.
2. What is the pH of a 0.15 M solution of CH_3COOK? K_a for $CH_3COOH = 1.8 \times 10^{-5}$.

(Answers: 1. 5.15; 2. 8.96)

Chapter 6. Buffers

This chapter will present problems to solve for the pH of solutions in which the "the common-ion effect" has occurred and for which a buffer condition exists.

Section 6.1 The Common Ion Effect

In these reactions, addition of a solute will cause a shift in the equilibrium. Thus, the pH of a weak acid or base in solution will change when that same acid or base is dissolved in an aqueous environment where a common ion exists. For example, HF in water hydrolyzes by the reaction:

$$HF\ (aq) + H_2O\ (l) \rightleftharpoons H_3O^+\ (aq) + F^{-1}\ (aq)$$

A common ion addition would be to place HF in a solution that contains F⁻ ions or is acidic. With the presence of F⁻ ions, this equilibrium shifts to the left; thus, less H_3O^+ is formed. Thus, the pH will numerically increase. In all cases, ICE tables will, again, be utilized to solve these problems.

Sample Problem 6.1 pH and the common ion effect

A 0.16 M solution of HF has a pH = 1.98, what is the pH of the same concentration (0.16M) of HF placed into a solution containing 0.05 M NaF? K_a for HF = 6.8 x 10⁻⁴.

Step 1. Identify the type of acid/base problem and the known and unknown variables.

This problem asks for the pH in a non-pure water solution; thus, it is a common ion effect problem. You know the concentration of The HF and NaF, and you must determine the pH.

A 0.16 M solution of HF has a pH = 1.98, what is the pH of the same concentration (0.16M) of HF placed into a solution containing 0.05 M NaF? K_a for HF = 6.8 x 10⁻⁴.

Step 2. Write the equilibrium expression, and create the ICE table.

$$HF\ (aq) + H_2O\ (l) \rightleftharpoons H_3O^+\ (aq) + F^{-1}(aq)$$

$$K_a = \frac{[H_3O^+][F^{-1}]}{[HF]}$$

	HF (aq)	+ H₂O (l)	⇌	H₃O⁺(aq) +	F⁻¹(aq)
Initial					
Change					
Equilibrium					

Step 3. Fill in the initial conditions, the change, and the equilibrium values.

Initially, there is 0.16 M HF and 0.05M NaF in solution. H₂O is in a liquid state and is excluded from the equilibrium expression.

	HF (aq)	+ H₂O (l)	⇌	H₃O⁺(aq) +	F⁻¹(aq)
Initial	0.16 M			0	0.05 M
Change					
Equilibrium					

Because H₃O⁺ has not formed yet, this equilibrium shifts right.

	HF (aq)	+ H₂O (l)	⇌	H₃O⁺(aq) +	F⁻¹(aq)
Initial	0.16 M			0	0.05 M
	SHIFT		→		
Change					
Equilibrium					

Assess the change, x.

Because the equilibrium shifts right, HF will deplete, and the other two components will form. Thus, the change in HF is negative, and the change in H₃O⁺ and F⁻¹ are positive.

	HF (aq)	+ H₂O (l)	⇌	H₃O⁺(aq) +	F⁻¹(aq)
Initial	0.16 M			0	0.05
	SHIFT		→		
Change	-x			+x	+x
Equilibrium					

Write the equilibrium condition.

	HF (aq)	+ H₂O (l) ⇌	H₃O⁺(aq) +	F⁻¹(aq)
Initial	0.16 M		0	0.05
	SHIFT	→		
Change	-x		+x	+x
Equilibrium	0.16 - x		+x	0.05+x

Step 4. Place the equilibrium values from the ICE table into the equilibrium expression, and solve for x.

$$K_a = \frac{[H_3O^+][F^{-1}]}{[HF]}$$

$$6.8 \times 10^{-4} = \frac{(x)(0.05+x)}{(0.16-x)}$$

Upon first inspection of this equation, a polynomial exists, and you can absolutely solve these problems can using the quadratic equation. Like in the prior examples, you can calculate the negligibility of x to determine if you can exclude it from both the expression for HF and F⁻.

Step 5. Assess x negligibility.

Use the following equation to see if x can be deemed negligible:

$$\frac{Initial\ Concentration\ of\ Weak\ Acid}{K_a\ of\ weak\ acid} > 100$$

For this problem:

$$\frac{0.16\ M}{6.8 \times 10^{-4}} > 100$$

$$235 > 100$$

Here 235 is greater than 100; thus, x is negligible, and you can exclude it from the equilibrium expression for both HF and F⁻.

Step 6. Solve for x, and calculate pH.

$$6.8 \times 10^{-4} = \frac{(x)(0.05+\cancel{x})}{(0.16-\cancel{x})}$$

$$6.8 \times 10^{-4} = \frac{(x)(.05)}{(0.16)}$$

$$1.1 \times 10^{-4} = (x)(0.05)$$

$$x = 2.2 \times 10^{-3}$$

In this problem, x = [H$_3$O$^+$] (weak acid hydrolysis); thus, the pH is:

$$pH = -\log[H_3O^+]$$

$$pH = -\log[2.2 \times 10^{-4}]$$

$$pH = 2.66$$

STUDENT PROBLEMS

1. What is the pH of a solution containing .55 M acetic acid, CH$_3$COOH, and 0.10 M sodium acetate, CH$_3$COONa? K$_a$ for acetic acid is 1.8×10^{-5}.
2. What is the pH of a solution of a solution containing 0.33 M ethylamine C$_2$H$_5$NH$_2$ and 0.033 M C$_2$H$_5$NH$_3$Cl? K$_b$ for ethylamine, C$_2$H$_5$NH$_2$ = 4.7×10^{-4}.
(Answers: 1. 4.00; 2. 11.67)

Section 6.2 Buffers

Section 6.2.1 Common Ion Effect and Henderson-Hasselbalch Equations

A buffer is a solution that can accept small amounts of a strong acid or base and the pH value of the buffer minimally changes. A buffer is a combination of a weak acid and its conjugate base or a weak base and a conjugate acid. For example, HF and NaF or NH_3 and NH_4Br in solution together form a buffer. A buffer is a common ion effect problem as presented in the previous section.

The alternative method to pH calculation in buffers is through the use of the Henderson-Hasselbalch Equation. This equation is:

$$\text{Acid buffers: pH} = pK_a + \log \frac{[conjugate\ base]}{[weak\ acid]}$$

and

$$\text{Basic Buffers: pOH} = pK_b + \log\left(\frac{[conjugate\ acid]}{[weak\ base]}\right)$$

A new term is now introduced pK_a and pK_b, and you can obtain these values through the use of these equations:

$$pK_a = -\log(K_a)$$

$$pK_b = -\log(K_b)$$

In this section, both methods will be used to solve these problems.

Sample Problem 6.2 pH of a buffer using common ion effect method

What is the pH of a buffer that contains 500.0 mL of 0.10 M NH_3 and 500.0 mL of 0.10 M NH_4Cl? The K_b for NH_3 equals 1.8×10^{-5}.

Step 1. Identify the type of acid/base problem and the known and unknown variables.

Since this question states that this is a buffer, you will use the common ion effect method. You know the concentrations of NH_3 and NH_4^+; you must determine the pH.

What is the pH of a buffer that contains 500.0 mL of 0.10 M NH_3 and 500.0 mL of 0.10 M NH_4Cl? The K_b for NH_3 equals 1.8×10^{-5}.

Note: In equilibria, you must utilize the concentration of the components; thus, you will not utilize volumes in the solution of this problem.

Step 2. Write the equilibrium expression, and create the ICE table.

$$NH_3 \, (aq) + H_2O \, (l) \rightleftharpoons OH^- \, (aq) + NH_4^{+1} \, (aq)$$

$$K_b = \frac{[NH_4^{+1}][OH^{-1}]}{[NH_3]}$$

	NH_3 (aq)	+ H_2O (l)	⇌	OH^- (aq)	+ NH_4^{+1} (aq)
Initial					
Change					
Equilibrium					

Step 3. Fill in the initial conditions, the change, and the equilibrium values.

Initially, there is 0.10 M NH_3 and 0.10 M NH_4Cl. Remember, H_2O is in a liquid sate and is excluded from the equilibrium expression.

	NH_3 (aq)	+ H_2O (l)	⇌	OH^- (aq)	+ NH_4^{+1} (aq)
Initial	0.10 M			0	0.10 M
Change					
Equilibrium					

Because OH^- has not formed yet, this equilibrium shifts right.

	NH$_3$ (aq)	+ H$_2$O (l) ⇌	OH$^-$ (aq)	+ NH$_4^{+1}$ (aq)
Initial	0.10 M		0	0.10 M
SHIFT		→		
Change				
Equilibrium				

Assess the change, x.

	NH$_3$ (aq)	+ H$_2$O (l) ⇌	OH$^-$ (aq)	+ NH$_4^{+1}$ (aq)
Initial	0.10 M		0	0.10 M
SHIFT		→		
Change	-x		+x	+x
Equilibrium				

Write the equilibrium condition.

	NH$_3$ (aq)	+ H$_2$O (l) ⇌	OH$^-$ (aq)	+ NH$_4^{+1}$ (aq)
Initial	0.10 M		0	0.10
SHIFT		→		
Change	-x		+x	+x
Equilibrium	0.10 - x		+x	0.10 + x

Once you have completed the table, you can now solve the problem for x.

Step 4. Place the equilibrium values from the ICE table into the equilibrium expression, and solve for x.

$$K_b = \frac{[NH_4^{+1}][OH^{-1}]}{[NH_3]}$$

$$1.8 \times 10^{-5} = \frac{(x)(0.10+x)}{(0.10-x)}$$

Upon first inspection of this equation, a polynomial exists, and you can absolutely solve this problem using the quadratic equation. Or, you can complete the assessment of negligibility of x.

Step 5. Assess x negligibility.

Use the following equation to see if x can be deemed negligible:

$$\frac{\text{Initial Concentration of Weak Base}}{K_b \text{ of weak base}} > 100$$

For this problem:

$$\frac{0.10 \text{ M}}{1.8 \times 10^{-5}} > 100$$

$$5556 > 100$$

5556 is greater than 100; thus, x is negligible and can be excluded from the equilibrium expression.

Step 6. Solve for x, and calculate pOH.

For weak base equilibria, the x in the expression stands for the [OH⁻]. Thus, you can determine the pOH then the pH.

$$1.8 \times 10^{-5} = \frac{(x)(0.10 + \cancel{x})}{(0.10 - \cancel{x})}$$

$$1.8 \times 10^{-5} = \frac{(x)(\cancel{0.10})}{(\cancel{0.10})}$$

$$1.8 \times 10^{-5} = x$$

In this problem, x = [OH⁻]; thus, the pOH is:

$$pOH = -\log[OH^-]$$

$$pOH = -\log[1.8 \times 10^{-5}]$$

$$pOH = 4.74$$

Thus, to solve for pH:

$$pH + pOH = 14.00$$

$$pH + 4.74 = 14.00$$

$$pH = 9.26$$

Sample Problem 6.3 pH of a buffer using the Henderson-Hasselbalch equation

Solving this same question with the Henderson-Hasselbalch equation:

What is the pH of a buffer that contains 500.0 mL of 0.10 M NH_3 and 500.0 mL of 0.10 M NH_4Cl? The K_b for NH_3 equals 1.8×10^{-5}.

Step 1. Identify the type of acid/base problem and the known and unknown variables.

Since this question states that this is a NH_3/NH_4^+ buffer, the base buffer form of the Henderson-Hasselbalch equation will be used. You know the concentrations of NH_3 and NH_4; you must determine the pH.

What is the pH of a buffer that contains 500.0 mL of 0.10 M NH_3 and 500.0 mL of 0.10 M NH_4Cl? The K_b for NH_3 equals 1.8×10^{-5}.

Note: In equilibria, you must utilize the concentration of the components; thus, you will not utilize volumes in the solution of this problem.

Step 2. Utilizing the Henderson-Hasselbalch equation, fill in the known variables, and solve for pH.

$$pOH = pK_b + \log\left(\frac{[conjugate\ acid]}{[weak\ base]}\right)$$

$$pOH = -\log(K_b) + \log\left(\frac{[conjugate\ acid]}{[weak\ base]}\right)$$

$$pOH = -\log(1.8 \times 10^{-5}) + \log\left(\frac{[0.10]}{[0.10]}\right)$$

$$pOH = 4.74$$

$$pH = 14.00 - pOH$$

$$pH = 14.00 - 4.74$$

$$pH = 9.26$$

STUDENT PROBLEMS

1. What is the pH of a buffer that contains 0.22 M HNO_2 and 0.33M $LiNO_2$?
   K_a for $HNO_2 = 4.5 \times 10^{-4}$
2. What is the pH of 100.0 mL solution that contains 500.0 moles of methylamine, CH_3NH_2, and 700.0 moles of methylammonium chloride, CH_3NH_3Cl? K_b of $CH_3NH_2 = 4.4 \times 10^{-4}$.
   (Answers: 1. 3.52; 2. 10.50)

Section 6.3 Calculations for the Addition of Acids or Bases to Buffers

As mentioned at the beginning of this chapter, a buffer is defined as a solution that can accept a small amount of a strong acid or base, and its pH will not dramatically change. These sets of problems show the methods utilized to calculate these pH values.

Section 6.3.1 Addition of a Strong Acid

Solving Tips

1. First, calculate either the number of moles or mmoles of the buffer constituents and the added acid.
2. Write the reaction that occurs in this solution. Note: the weak acid in the buffer reacts with a strong base to form more conjugate base. Alternatively, the conjugate base in the buffer will react with the addition of a strong acid, forming more weak acid.

For example:

There is a NH_3/NH_4^+ buffer system. The reaction that occurs when HCl is added is:

$$NH_3 \text{ (aq)} + HCl \rightleftharpoons NH_4^+ \text{ (aq)} + Cl^- \text{ (aq)}$$

Base — Acid — conjugate acid

If NaOH, a strong base, is added to this same buffer, this reaction occurs:

$$NH_4^+ \text{ (aq)} + NaOH \text{ (aq)} \rightleftharpoons NH_3 \text{ (aq)}$$

Conjugate Acid — Base — weak base

Note: As it was stated in Chapter 5, Acids and Bases, the weak acid and its conjugate base or the weak base and its conjugate acid are forever linked in these reactions.

3. Once the new amounts of weak base and conjugate are known, utilize the Henderson-Hasselbalch equation to solve.
4. When using the Henderson-Hasselbalch equation, make sure to correctly fill in the correct amounts for the weak acid or base and their conjugate. This is a very common mistake to flip the ratio.

Sample Problem 6.4 pH of a buffer after addition of strong acid

A buffer contains 500.0 mL of 0.100 M NH_3 and 0.100 M NH_4Cl? What is the pH of this buffer after the addition of 100.0 mL of 0.300 M HCl? The K_b for NH_3 equals 1.8×10^{-5}.

Step 1. Identify the type of acid/base problem and the known and unknown variables.

This is a buffer problem, as stated in the question itself. Thus, either the common ion effect or Henderson-Hasselbalch equation can be used to solve this problem.

A buffer contains 500.0 mL of 0.100 M NH_3 and 0.100 M NH_4Cl? What is the pH of this buffer after the addition of 100.0 mL of 0.300 M HCl? The K_b for NH_3 equals 1.8×10^{-5}.

The volume of the buffer, the concentration of NH_3 and NH_4^+, and the amount of HCl added are known; the pH is unknown.

Step 2. Calculate the initial amounts of NH_3, NH_4^+, and HCl.

$$\text{mmoles} = M \times mL$$

Initial amounts of each component:

$$\text{mmoles of } NH_3 = 500.0 \text{ mL} \times 0.100 \text{ M} = 50.0 \text{ mmole}$$

$$\text{mmoles of } NH_4^{+1} = 500.0 \text{ mL} \times 0.100 \text{ M} = 50.0 \text{ mmole}$$

$$\text{mmoles HCl} = 100.0 \text{ mL} \times 0.300 \text{ M} = 30.0 \text{ mmole}$$

Step 3. Calculate the post-reaction amounts of NH_3, NH_4^+, and HCl.

The reaction that will occur when HCl is added is:

$$NH_3 \text{ (aq)} + HCl \text{ (aq)} \rightleftharpoons NH_4^+ \text{ (aq)} + Cl^- \text{ (aq)}$$

Create a table to calculate the final amounts:

Initial

	NH_3 (aq)	HCl (aq)	⇌	NH_4^+ (aq) +	Cl^- (aq)
Initial	50.0 mmol	30.0 mmol		50.0 mmol	
Reacted					
Final					

Chapter 6. Buffers | 127

To determine the amount reacted, one must compare the initial amounts of both reactants. Since a 1:1 stoichiometry exists in this reaction, the maximum amount that can react for both HCl and NH_3 is 30 mmol.

	NH_3 (aq)	HCl (aq)	⇌	NH_4^+ (aq) +	Cl^- (aq)
Initial	50.0 mmol	30.0 mmol		50.0 mmol	
Reacted	-30.0 mmol	-30.0 mmol		+30.0 mmol	
Final					

Final Amounts:

	NH_3 (aq)	HCl (aq)	⇌	NH_4^+ (aq) +	Cl^- (aq)
Initial	50.0 mmol	30.0 mmol		50.0 mmol	
Reacted	-30.0 mmol	-30.0 mmol		+30.0 mmol	
Final	20.0 mmol	0.0 mmol		80.0 mmol	

These final mmole amounts are then placed into the Henderson-Hasselbalch equation to solve for the pH.

Note: In this equation, a ratio of the conjugate acid exists to the weak base. You do not have to convert these values back to molarity, because the components are in the same volume.

$$\frac{[conjugate\ acid]}{weak\ base}$$

$$\frac{\frac{80.0\ mmol}{500.0\ mL}}{\frac{20.0\ mmol}{500.0\ mL}}$$

$$\frac{80.0}{500.0} \times \frac{500.0}{20.0}$$

$$\frac{80.0\ mmol}{20.0\ mmol}$$

Step 4. Solve for the pH utilizing the Henderson-Hasselbalch equation.

$$pOH = pK_b + \log\left(\frac{[conjugate\ acid]}{[weak\ base]}\right)$$

$$pOH = -\log(K_b) + \log\left(\frac{[conjugate\ acid]}{[weak\ base]}\right)$$

$$pOH = -\log(1.8 \times 10^{-5}) + \log\left(\frac{(80.0)}{(20.0)}\right)$$

$$pOH = 5.34$$

$$pH = 14 - pOH$$

$$pH = 14 - 5.34$$

$$pH = 8.66$$

Recall, this same buffer from the prior problem had a pH of 9.26 before any addition of acid. Adding the strong acid numerical dropped the pH to 8.66.

Section 6.3.2 Addition of a Strong Base

In this scenario, NaOH, a strong base, is added to the NH_3/NH_4^+ buffer, and this reaction occurs:

$$NH_4^+\ (aq) + NaOH\ (aq) \rightleftarrows NH_3\ (aq)$$

Conjugate Acid   Base   weak base

Sample Problem 6.5 pH of a buffer after addition of strong base

A buffer contains 500.0 mL of 0.10 M NH_3 and 0.10 M NH_4Cl? What is the pH of this buffer after the addition of 100.0 mL of 0.25 M NaOH? The K_b for NH_3 equals 1.8×10^{-5}.

Step 1. Identify the type of acid/base problem and the known and unknown variables.

This is a buffer problem as stated in the question itself. Thus, you can use either the common ion effect or Henderson-Hasselbalch equation to solve this problem.

A buffer contains 500.0 mL of 0.10 M NH_3 and 0.10 M NH_4Cl? What is the pH of this buffer after the addition of 100.0 mL of 0.25 M NaOH? The K_b for NH_3 equals 1.8×10^{-5}.

You know the volume of the buffer, the concentration of NH_3 and NH_4^+, and the amount of HCl added; you do not know the pH.

Step 2. Calculate the initial amounts of NH_3, NH_4^+, and NaOH.

$$\text{mmoles} = M \times mL$$

Initial amounts of each component:

$$\text{mmoles of } NH_3 = 500.0 \text{ mL} \times 0.10 \text{ M} = 50.0 \text{ mmole}$$

$$\text{mmoles of } NH_4^{+1} = 500.0 \text{ mL} \times 0.10 \text{ M} = 50.0 \text{ mmole}$$

$$\text{mmoles NaOH} = 100.0 \text{ mL} \times 0.25 \text{ M} = 25.0 \text{ mmole}$$

Step 3. Calculate the post-reaction amounts of NH_3, NH_4^+, and HCl.

The reaction that will occur when NaOH is added is:

$$NH_4^+ \text{ (aq)} + NaOH \text{ (aq)} \rightleftarrows NH_3 \text{ (aq)}$$

Create a table to calculate the final amounts:

Initial

	NH_4^+ (aq) +	NaOH (aq)	⇌	NH_3 (aq)
Initial	50.0 mmol	25.0 mmol		50.0 mmol
Reacted				
Final				

To determine the amount reacted, one must compare the initial amounts of both reactants. Since a 1:1 stoichiometry exists in this reaction, the maximum amount that can react for both NaOH and the NH_4^{+1} is 25 mmol.

	NH_4^+ (aq) +	NaOH (aq)	⇌	NH_3 (aq)
Initial	50.0 mmol	25.0 mmol		50.0 mmol
Reacted	-25.0 mmol	-25.0 mmol		+25.0 mmol
Final				

Final Amounts:

	NH_4^+ (aq) +	NaOH (aq)	⇌	NH_3 (aq)
Initial	50.0 mmol	25.0 mmol		50.0 mmol
Reacted	-25.0 mmol	-25.0 mmol		+25.0 mmol
Final	25.0 mmol	0.0 mmol		75.0 mmol

Step 4. Solve for the pH utilizing the Henderson-Hasselbalch equation.

$$pOH = pK_b + \log(\frac{[conjugate\ acid]}{[weak\ base]})$$

$$pOH = -\log(K_b) + \log(\frac{[conjugate\ acid]}{[weak\ base]})$$

$$pOH = -\log(1.8 \times 10^{-5}) + \log(\frac{(25)}{(75)})$$

$$pOH = 4.26$$

$$pH = 14 - pOH$$

$$pH = 14 - 4.26$$

$$pH = 9.74$$

Recall, that this same buffer from the prior problem had a pH of 9.26 before the strong base was added. Adding the strong base numerical raised the pH to 9.74.

STUDENT PROBLEM

1. What is the pH of a 600mL buffer that contains 0.45 M HF and 0.65 M NaF? K_a for HF = 3.5×10^{-4}.

   1a. What is the pH of this buffer after 0.090 moles of HCl are added?
   1b. What is the pH of this buffer after 150.0 mL of 0.1 M NaOH are added?

(Answers: 1. 3.38; 2. 3.67)

Section 6.4 Forming a Buffer with a Specific pH

The determination of the ratio of the buffer constituents (conjugate base to weak acid or conjugate acid to weak base) is found when a specific pH of this buffer solution is desired. Again, these types of problems are commonly solved with the Henderson-Hasselbalch equation.

Sample Problem 6.6 Forming a buffer with a specific pH

What quantity of weak acid and conjugate base are needed to create a buffer with a pH = 3.2?

Step 1. Identify the type of acid/base problem and the known and unknown variables.

<u>What quantity of weak acid and conjugate base</u> are needed to create a buffer with a pH = 3.2?

A buffer solution exists; thus, you can use the Henderson-Hasselbalch equation. You know the pH; however, you do not know the ratio of conjugate base to weak acid.

You will also need the pK_a values for the weak acids to solve this problem. These are commonly found in the textbook, and a short list is:

Table 6.1 Weak Acid K_a and pKa Values

Weak Acid	K_a value	$pK_a = -\log(K_a)$
HCN	4.9×10^{-10}	9.31
Benzoic Acid	6.5×10^{-5}	4.18
Formic Acid	1.8×10^{-4}	3.74
HF	3.5×10^{-4}	3.45
HNO_2	4.6×10^{-4}	3.33
CH_3COOH	1.8×10^{-5}	4.74

Step 2. Select a weak acid with a pK_a value close to the desired pH. Fill the variable into the Henderson-Hasselbalch equation, and solve.

Upon inspection of the pK_a values, HNO_2 has the value closest to the desired 3.20 amount.

$$pH = pK_a + \log \frac{[conjugate\ base]}{[weak\ acid]}$$

$$3.20 = 3.33 + \log \frac{[conjugate\ base]}{[weak\ acid]}$$

$$-0.13 = \log \frac{[\text{conjugate base}]}{[\text{weak acid}]}$$

Take the anti-log of both sides to get:

$$10^{-.13} = \frac{[\text{conjugate base}]}{[\text{weak acid}]}$$

$$0.741 = \frac{[\text{conjugate base}]}{[\text{weak acid}]} = \frac{0.741}{1}$$

Thus mixing 1 mole of HNO_2 in solution with 0.741 moles of its conjugate base NO_2^- creates a buffer with a pH = 3.20

Note: the ratio cannot exceed the value of 10.

STUDENT PROBLEMS

1. What quantity of weak acid and conjugate base do you need to create a buffer with a pH = 3.90?
2. What quantity of weak base and conjugate acid do you need to create a buffer with a pH = 8.88?

(Answers: 1. 1.45:1; 2. 0.372:1)

Chapter 7. Titrations

This chapter will show how to solve acid/base problems based on the wet chemical method called titration. You commonly use titration to determine the concentration or gram quantity of an unknown substance in solution. The titrant is the solution in the buret, and the unknown substance is in the titration flask. The four other columns show what condition exists at different points in the titration. This chapter will present problems where you will determine the pH at various points in the titration. The tables below list the three different types of titration that you will explore.

Note: In column one, the top solution is in the titration flask; the bottom solution is the titrant

You can use these tables to help solve these problems:

Table 7.1 Types of Acid/Base Titrations and Resulting Conditions

Type of Titration	Starting Point (no titrant)	Before Equivalence Point	Equivalence Point	After Equivalence Point
Strong Acid Strong Base	Strong Acid	Strong Acid	Neutral	Strong Base
Weak Acid Strong Base	Weak Acid Hydrolysis	Acid Buffer	Salt Solution K_b – Conjugate Base	Strong Base
Weak Base Strong Acid	Weak Base Hydrolysis	Basic Buffer	Salt Solution K_a – Conjugate Acid	Strong Acid

Table 7.2 Required Equations for Any Point in Acid/Base Titrations

Type of Titration	Starting Point (no titrant)	Before Equivalence Point	Equivalence Point	After Equivalence Point
Strong Acid Strong Base	$pH = -\log[H_3O^+]$	$pH = -\log[H_3O^+]$	7	$pOH = -\log[OH^-]$
Weak Acid Strong Base	$K_a = x^2/[\text{weak acid}]$ $x = [H_3O^+]$ $pH = -\log[H_3O^+]$	$pH = pK_a + \log\dfrac{[\text{conjugate base}]}{[\text{weak acid}]}$	$K_w/K_a = x^2/[\text{conj base}]$ $x = [OH^-]$ $pOH = -\log[OH^-]$	$pOH = -\log[OH^-]$
Weak Base Strong Acid	$K_b = x^2/[\text{weak base}]_i$ $x = [OH^-]$ $pOH = -\log[OH^-]$ $pH = 14 - pOH$	$pOH = pK_b + \log\dfrac{[\text{conjugate acid}]}{[\text{weak base}]}$	$K_w/K_b = x^2/[\text{conj acid}]$ $X = [H_3O^+]$ $pH = -\log[H_3O^+]$	$pH = -\log[H_3O^+]$

Solving Tips

1. Always look to see what type of acid or base the problem presents. This will allow you to identify the type of titration being performed.
2. Calculate the amount of titrant and flask contents first. This will allow for the immediate identification of the point you are solving for in the titration.
3. Remember, you are measuring or calculating the pH of the solution in the flask.
4. Work in mmole and mL to avoid conversion errors.
5. To correctly calculate the pH of the resulting solution, remember to calculate the final concentration with the TOTAL VOLUME in the flask.
6. When solving titration problems that involve a weak and a strong acid or base, the amount of the component that limits in the reaction will be the amount of conjugate acid or base formed.
7. After the equivalence point in the weak acid/strong base titrations, both a strong base [OH⁻] and conjugate base [A⁻] are present. And both can contribute to the pH of the solution. However, the [OH⁻] overwhelms the amount of A⁻ present. Thus, the A⁻ contribution is negligible compared to the pH effect of the OH⁻. This will also be the case for a weak base/strong acid titration. The concentration of strong acid dictates the pH after the equivalence point.

Section 7.1 Strong Acid/Strong Base Titrations

Since these types of titrations are solely strong acids and bases, the set of equations needed for solving are:

$$pH = -\log[H_3O^+]$$

$$pOH = -\log[OH^-]$$

$$pH + pOH = 14$$

Section 7.1.1 Initial State – No Titrant – Strong Acid/Strong Base

Sample Problem 7.1 pH of the initial state – no titrant – strong acid/strong base

What is the pH of 25.0 mL of a 0.05 M solution of HCl before the addition of the titrant 0.05 M NaOH?

Step 1. Identify the type of acid/base titration, the flask solution, and the titrant.

In these problems, an acid/base or neutralization reaction occurs. The key is determining which reactant is the limiting reagent. For this problem, HCl and NaOH are reactants and are a STRONG Acid and STRONG base. Thus, the solution will follow this row from the chart.

Titration Type	Starting Point (no titrant)	Before Equivalence Point	Equivalence Point	After Equivalence Point
Strong Acid Strong Base	$pH = -\log[H_3O^+]$	$pH = -\log[H_3O^+]$	7	$pOH = -\log[OH^-]$

As shown, the HCl serves as the flask solution, and as the problem states, the NaOH serves as the titrant.

Step 2. Use the stacking method to identify the quantity of each component.

Multiply the volume amount of each component by their concentration to yield the mmole values of each:

Flask	HCl	25.0 mL	x	0.05 M	=	1.25 mmol HCl
Titrant	NaOH	0.0 mL	x	0.05 M	=	0.0 mmol NaOH

Since there is no NaOH in the flask, the only component contributing to the pH is the HCl.

Step 3. Solve for pH of the solution in the flask.

The flask contains 25.0 mL of 0.05 M HCl, and you only need the concentration at the initial point of the titration to solve for the pH. Thus:

$$pH = -\log[H_3O^+]$$

$$pH = -\log [0.05 \text{ M}]$$

$$pH = 1.30$$

Section 7.1.2 Before the Equivalence Point – Strong Acid/Strong Base

NOTE: In these problems, the titrant (strong base) limits in the reaction, and HCl remains.

Sample Problem 7.2 pH before the equivalence point – strong acid/strong base

What is the pH of a solution of 25.0 mL of a 0.05 M solution of HCl after 10.0 mL of 0.05 M NaOH has been added?

Step 1. Identify the type of acid/base titration, the flask solution, and the titrant.

This is a strong acid (HCl)/strong base (NaOH) problem. The HCl is in the flask, and the NaOH is the titrant.

<u>What is the pH</u> of a solution of ⬭25.0 mL of a 0.05 M solution of HCl after 10.0 mL of 0.05 M NaOH⬭ has been added?

          You know NaOH is the titrant because of the word "after" and "has been added"

Step 2. Use the stacking method to identify the quantity of each component.

Multiply the volume of each component by their concentration to yield the mmole values of each:

Flask	HCl	25.0 mL	x	0.05 M	=	1.25 mmol HCl
Titrant	NaOH	10.0 mL	x	0.05 M	=	0.5 mmol NaOH

There are two ways to approach this next step, write the neutralization reaction with initial, changing, and resulting values or simply subtract.

Step 2a. Use the neutralization reaction.

In the neutralization reaction, compare the quantities of HCl and NaOH. In this problem, the NaOH is smaller and, thus, limits.

$$HCl\ (aq) + NaOH\ (aq) \rightarrow H_2O\ (l) + NaCl\ (aq)$$

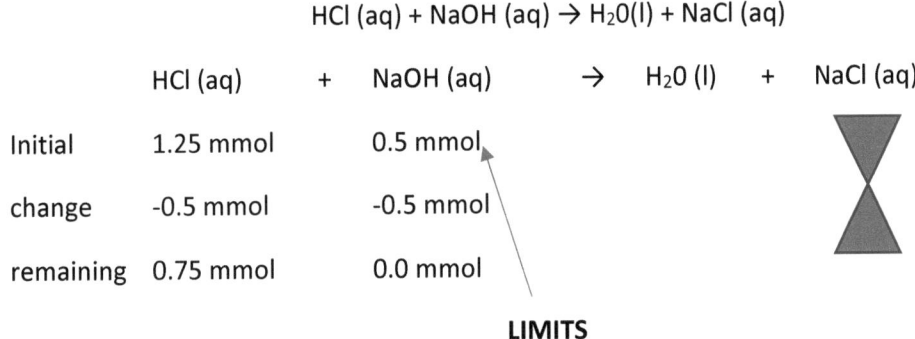

LIMITS

Or Step 2b. Simply subtract.

Using the first table:

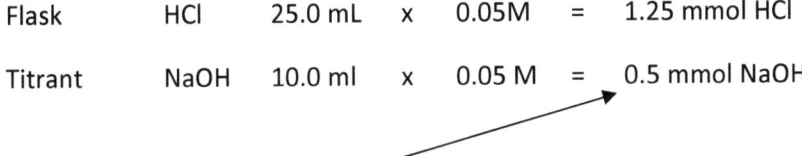

Subtract the values in the last column and assess what remains:

1.25 mmol HCl

-0.5 mmol NaOH

0.75 mmol HCl remaining

Step 3. Calculate the TOTAL VOLUME in the flask.

From the table:

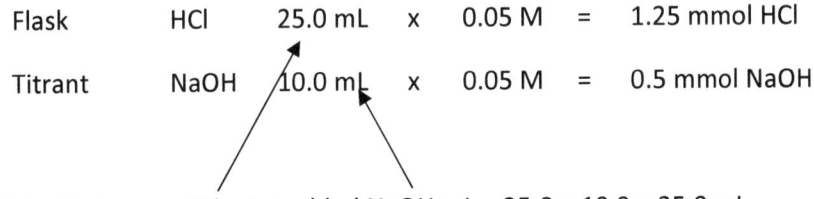

Total Volume = HCl mL + added NaOH mL = 25.0 + 10.0 = 35.0 mL

Step 4. Calculate the pH of the resulting flask solution.

The equation to solve for pH is:

$$pH = -\log[H_3O^+]$$

Thus, you will need the CONCENTRATION, and it is the ratio of mmol of HCl in step 2 (0.75 mmol) to the total volume in Step 3 (35.0 mL). The H_3O^+ concentration is the resulting concentration of the HCl, which is:

$$\frac{mmol\ HCl}{Total\ Volume\ in\ mL}$$

$$\frac{0.75\ mmol\ HCl}{35.0\ mL} = 0.021\ M$$

$$pH = -\log(0.021)$$

$$pH = 1.67$$

Section 7.1.3 At the Equivalence Point – Strong Acid/Strong Base

The equivalence point of a titration is where the mmoles of the flask solution are equal to the mmoles of the titrant.

Sample Problem 7.3 pH at the equivalence point – strong acid/strong base

What is the pH of a solution of 25.0 mL of a 0.05 M solution of HCl after 25.0 mL of 0.05 M NaOH has been added?

Step 1. Identify the type of acid/base titration, the flask solution, and the titrant.

This serves as a strong acid (HCl)/strong base (NaOH) problem. The HCl is in the flask, and the NaOH serves as the titrant.

<u>What is the pH</u> of a solution of 25.0 mL of a 0.05 M solution of HCl after 25.0 mL of 0.05M NaOH has been added?

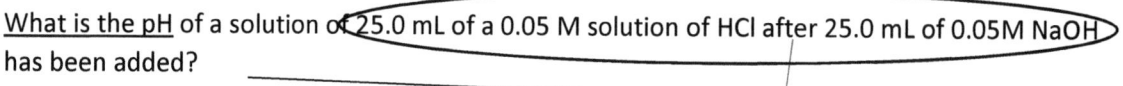

You know NaOH is the titrant because of the word "after" and "has been added"

Step 2. Use the stacking method to identify the quantity of each component.

Multiply the volume amount of each component by their concentration to yield the mmole values of each:

| Flask | HCl | 25.0 mL | x | 0.05 M | = | 1.25 mmol HCl |
| Titrant | NaOH | 25.0 mL | x | 0.05 M | = | 1.25 mmol NaOH |

Again, you can approach this next step in two ways: write the neutralization reaction with initial, changing, and resulting values, or simply subtract.

Step 2a. Use the neutralization reaction.

In the neutralization reaction, compare the quantities of HCl and NaOH. In this problem, the NaOH amount is equal to the amount of HCl, which means the reaction is at the equivalence point.

$$HCl\ (aq) + NaOH\ (aq) \rightarrow H_2O\ (l) + NaCl\ (aq)$$

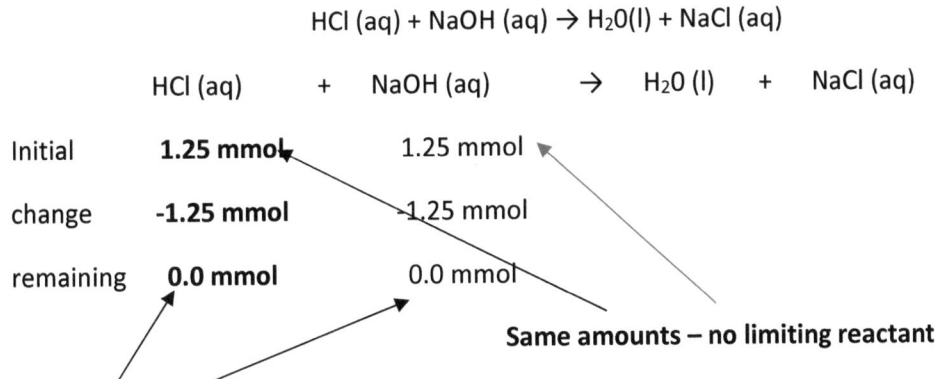

No HCl nor NaOH remain in solution

Same amounts – no limiting reactant

Or Step 2b. Simply subtract.

Using the first table:

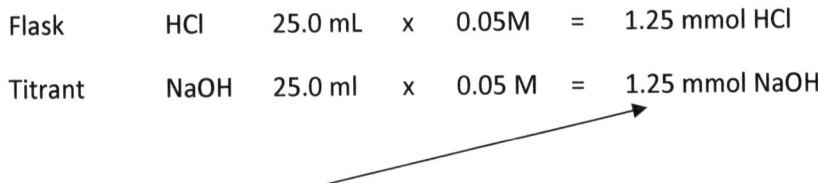

| Flask | HCl | 25.0 mL | x | 0.05M | = | 1.25 mmol HCl |
| Titrant | NaOH | 25.0 ml | x | 0.05 M | = | 1.25 mmol NaOH |

Subtract the values in the last column and assess what remains. Thus,

1.25 mmol HCl

<u>-1.25 mmol NaOH</u>

0.0 mmol of either HCl or NaOH remain

Step 4. Calculate the pH of the resulting flask solution.

In this case, neither HCl nor NaOH remain, or their values are equal. The resulting pH is then $[H_3O^+] = [OH^-] = 1.0 \times 10^{-7}$.

$$pH = -\log(1.0 \times 10^{-7}) = 7$$

Section 7.1.4 After the Equivalence Point – Strong Acid/Strong Base

At this point in the titration, excess titrant or NaOH remains.

Sample Problem 7.4 pH after the Equivalence Point – Strong Acid/Strong Base

What is the pH of a solution of 25.0 mL of a 0.05 M solution of HCl after 40.0 mL of 0.05 M NaOH has been added?

Step 1. Identify the type of acid/base titration, the flask solution, and the titrant.

This is a strong acid (HCl)/strong base (NaOH) problem. The HCl is in the flask, and the NaOH serves as the titrant.

What is the pH of a solution of 25.0 mL of a 0.05 M solution of HCl after 40.0 mL of 0.05 M NaOH has been added?

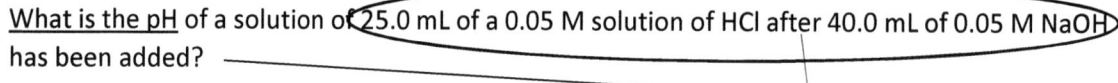

You know NaOH is the titrant because of the word "after" and "has been added"

Step 2. Use the stacking method to identify the quantity of each component.

Multiply the volume amount of each component by their concentration to yield the mmole values of each:

Flask	HCl	25.0 mL	×	0.05 M	=	1.25 mmol HCl
Titrant	NaOH	40.0 mL	×	0.05 M	=	2.0 mmol NaOH

Again, you can approach this next step in two ways, write the neutralization reaction with initial, changing, and resulting values or simply subtract.

Step 2a. Use the neutralization reaction.

In the neutralization reaction, compare the quantities of HCl and NaOH. In this problem, the HCl is smaller and, thus, limits.

$$HCl\ (aq) + NaOH\ (aq) \rightarrow H_2O(l) + NaCl\ (aq)$$

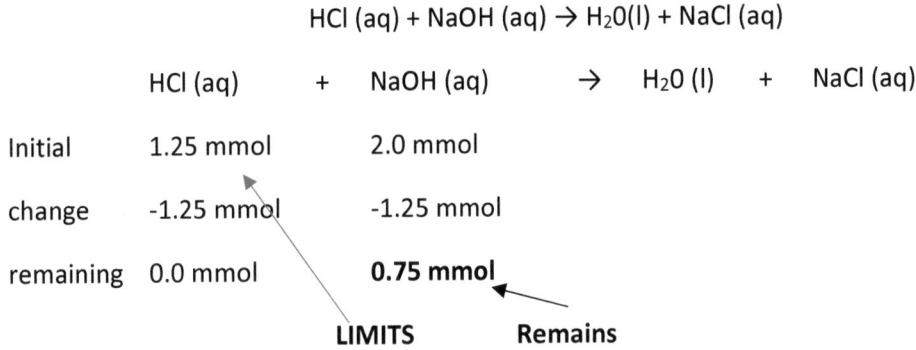

Or Step 2b. Simply subtract.

Using the first table:

| Flask | HCl | 25.0 mL | × | 0.05M | = | 1.25 mmol HCl |
| Titrant | NaOH | 40.0 ml | × | 0.05 M | = | 2.0 mmol NaOH |

Subtract the values in the last column and assess what remains. Thus,

1.25 mmol HCl
-2.0 mmol NaOH
0.75 mmol NaOH remain

Step 3. Calculate the TOTAL VOLUME in the Flask.

From the table:

| Flask | HCl | 25.0 mL | × | 0.05 M | = | 1.25 mmol HCl |
| Titrant | NaOH | 40.0 mL | × | 0.05 M | = | 2.0 mmol NaOH |

Total Volume = HCl mL + added NaOH mL = 25.0 + 40.0 = 65.0 mL

Step 4. Calculate the pH of the resulting flask solution.

Since the NaOH remains, solve for the pOH initially and then pH:

$$pOH = -\log[OH^-]$$

$$pH + pOH = 14$$

You must determine the CONCENTRATION and the ratio of mmol of NaOH in Step 2 (0.75 mmol) to the total volume in Step 3 (65.0 mL). The OH⁻ concentration is the resulting concentration of the NaOH, which is:

$$\frac{mmol\ NaOH}{Total\ Volume\ in\ mL}$$

$$\frac{0.75\ mmol\ NaOH}{65.0\ mL} = 0.0115\ M\ NaOH$$

pOH = - log (0.0115)

pOH = 1.94

pH = 14 − pOH

pH = 14 − 1.94 = 12.06

Section 7.2 Weak Acid/Strong Base Titrations

In these types of titrations, the weak acid is in the flask, and the strong base serves as the titrant. The equations needed to solve these problems are:

Starting Point: Weak Acid Hydrolysis: $K_a = \dfrac{x^2}{[weak\ acid]_{initial}}$; $x = [H_3O^+]$

Before Equivalence: Buffer: $pH = pK_a + \log \dfrac{[conjugate\ base]}{[weak\ acid]}$

At Equivalence Point: Conjugate Base Hydrolysis: $\dfrac{K_w}{K_a} = \dfrac{x^2}{[conjugate\ base]}$; $x = [OH^-]$

After Equivalence Point: Strong Base in Excess: $pOH = -\log(OH^-)$

Section 7.2.1 Initial State – No Titrant – Weak Acid/Strong Base

Sample Problem 7.5 pH of the initial state – no titrant – weak Acid/Strong base

What is the pH of 25.0 mL of a 0.10 M solution of HCN **before** the addition of the titrant 0.05 M NaOH? Ka for HCN = 4.9 x 10^{-10}.

Step 1. Identify the type of acid/base titration, the flask solution, and the titrant.

In these problems, an acid/base or neutralization reaction occurs. The key involves determining which reactant serves as the limiting reagent. For this problem, HCN and NaOH are reactants and are a WEAK Acid (K_a is given) and STRONG base. Thus, solution will follow this row from the chart:

Type of Titration	Starting Point (no titrant)	Before Equivalence Point	Equivalence Point	After Equivalence Point
Weak Acid Strong Base	$K_a = x^2/[weak\ acid]_i$ $X = [H_3O^+]$ $pH = -\log[H_3O^+]$	$pH = pK_a + \log\dfrac{[conjugate\ base]}{[weak\ acid]}$	$K_w/K_a = x^2/[conj\ base]$ $X = [OH-]$ $pOH = -\log[OH^-]$	$pOH = -\log[OH^-]$

The HCN serves as the flask solution, and as the problem states, the NaOH serves as the titrant.

Step 2. Use the stacking method to identify the quantity of each component.

Multiply the volume amount of each component by their concentration to yield the mmole values of each:

Flask	HCN	25.0 mL	x	0.10 M	=	2.5 mmol HCN
Titrant	NaOH	0.0 mL	x	0.05 M	=	0.0 mmol NaOH

Since there is no NaOH in the flask, the only component contributing to the pH is the HCN.

Step 3. Solve for pH of the solution in the flask.

There is solely 25.0 mL of 0.10 M HCN in the flask, which is a weak acid. To get the pH of this solution, as learned in the previous chapter, the equilibrium hydrolysis of HCN occurs by the reaction:

$$HCN\ (aq) + H_2O\ (l) \rightleftarrows H_3O^+\ (aq) + CN^-\ (aq)$$

Using an ICE table:

$$K_a = \frac{[H_3O^+][CN^{-1}]}{[HCN]}$$

	HCN (aq)	+ H$_2$O (l)	$\rightleftarrows$	H$_3$O$^+$(aq) +	CN^{-1}(aq)
Initial					
Change					
Equilibrium					

Write the initial conditions, the change, and the equilibrium condition.

	HCN (aq)	+ H$_2$O (l)	$\rightleftarrows$	H$_3$O$^+$(aq) +	CN^{-1}(aq)
Initial	0.10 M			0	0
	SHIFT		→		
Change	-x			+x	+x
Equilibrium	0.10 - x			+x	+x

Place the equilibrium values from the ICE table into the equilibrium expression, and solve for x.

$$K_a = \frac{[H_3O^+][CN^{-1}]}{[HCN]}$$

$$4.9 \times 10^{-10} = \frac{(x)(x)}{(0.10-x)}$$

Upon first inspection of this equation, a polynomial exists, and you can absolutely solve these problems using the quadratic equation. Like in the prior examples, you can calculate the negligibility of x to determine if you can exclude it from both the expression for HCN.

Step 4. Assess x negligibility.

Use the following equation to see if you can deem x negligible:

$$\frac{Initial\ Concentration\ of\ Weak\ Acid}{K_a\ of\ weak\ acid} > 100$$

For this problem:

$$\frac{0.10\ M}{4.9 \times 10^{-10}} > 100$$

$$2.0 \times 10^8 > 100$$

2.0×10^8 is greater than 100; thus, x is negligible and can be excluded from the equilibrium expression for HCN.

Step 5. Solve for x, and calculate pH.

$$4.9 \times 10^{-10} = \frac{x^2}{(0.10)}$$

$$x^2 = 4.9 \times 10^{-11}$$

$$\sqrt{x^2} = \sqrt{4.9 \times 10^{-11}}$$

$$x = 7.0 \times 10^{-6}$$

In this problem, x = [H$_3$O$^+$] (weak acid hydrolysis); thus, the pH is:

$$pH = -\log[H_3O^+]$$

$$pH = -\log[7.0 \times 10^{-6}]$$

$$pH = 5.15$$

Section 7.2.2 Before the Equivalence Point – Weak Acid/Strong Base

NOTE: In a weak acid/strong base titration, the base limits the reaction before the equivalence point. In this case, both the weak acid and its conjugate exist in the solution, which means a buffer forms.

$$HCN\ (aq) + NaOH\ (aq) \rightleftharpoons H_2O\ (l) + NaCN^-\ (aq)$$

    remains   limits   forms

The Na⁺ can be excluded from the equilibrium, because it is a weak conjugate that will not hydrolyze and will not affect the pH.

Sample Problem 7.6 pH before the equivalence point – weak acid/strong base

What is the pH of a solution of 25.0 mL of a 0.10 M solution of HCN after 15.0 mL of 0.05M NaOH has been added?

Step 1. Identify the type of acid/base titration, the flask solution, and the titrant.

This is a weak acid (HCN)/strong base (NaOH) problem. The HCN is in the flask, and the NaOH serves as the titrant.

What is the pH of a solution of 25.0 mL of a 0.10 M solution of HCN after 15.0 mL of 0.05M NaOH has been added? K_a of HCN = 4.9 x 10⁻¹⁰

    You know NaOH is the titrant because of the word "after" and "has been added"

Step 2. Use the stacking method to identify the quantity of each component.

Multiply the volume amount of each component by its concentration to yield the mmole values of each:

Flask HCN 25.0 mL x 0.10 M = 2.5 mmol HCN

Titrant NaOH 15.0 mL x 0.05 M = 0.75 mmol NaOH

Step 2a. Use the neutralization reaction.

In the neutralization reaction, compare the quantities of HCN and NaOH. In this problem, the NaOH is smaller and, thus, limits.

$$HCN\ (aq) + NaOH\ (aq) \rightleftharpoons H_2O\ (l) + CN^-\ (aq)$$

Chapter 7. Titrations | 147

	HCN (aq)	+	NaOH (aq)	→	H₂O (l)	+	CN⁻ (aq)
Initial	2.5 mmol		0.75 mmol				0.0 mmols
change	-0.75 mmol		-0.75 mmol		SHIFT →		+0.75 mmoles
remaining	1.75 mmol		0.0 mmol				0.75 mmoles
	REMAINS				LIMITS		FORMS

Or Step 2b. Use the subtraction method.

Using this table:

Flask	HCN	25.0 mL	×	0.10 M	=	2.5 mmol HCN
Titrant	NaOH	15.0 mL	×	0.05 M	=	0.75 mmol NaOH

Again, looking solely at the last column, you can simply determine the amount of HCN remaining and the amount of CN⁻ formed:

2.5 mmol HCN

0.75 mmol NaOH ← Limits = amount of CN⁻ produced

Subtract Values Yields Amount of HCN Remaining

2.5 mmol HCN

-0.75 mmol NaOH

1.75 mmol HCN

Since HCN and its conjugate CN⁻ coexist in solution; this is a buffer, and you can use the Henderson-Hasselbalch equation to solve for pH.

Step 3. Calculate the TOTAL VOLUME in the Flask.

This step is not needed for buffer solutions, because the weak acid and its conjugate are in the same volume. The mmole values of each will be ratioed to provide the given result.

Step 4. Calculate the pH of the resulting buffer solution in the flask.

Use the Henderson-Hasselbalch equation to solve for pH:

$$pH = pK_a + \log \frac{[conjugate\ base]_{formed}}{[weak\ acid]_{remaining}}$$

$$pH = -\log(4.9 \times 10^{-10}) + \log \frac{(0.75 \, mmol)}{(1.75 \, mmol)}$$

$$pH = 9.31 + (-0.368)$$

$$pH = 8.94$$

Section 7.2.3 At the Equivalence Point – Weak Acid/Strong Base

The equivalence point of a titration is where the mmoles of the flask solution are equal to the mmoles of the titrant. The set of equations that you will utilize to solve at this point in the titration are:

$$\frac{K_w}{K_a} = \frac{x^2}{[conjugate\,base]}$$

$$x = [OH^-]$$

$$pOH = -\log[OH^-]$$

$$pH = 14 - pOH$$

Sample Problem 7.7 pH at the equivalence point – weak acid/strong base

What is the pH of a solution of 25.0 mL of a 0.10 M solution of HCN after 50.0 mL of 0.05 M NaOH has been added? K_a of HCN = 4.9×10^{-10}.

Step 1. Identify the type of acid/base titration, the flask solution, and the titrant.

This is a weak acid (HCN)/strong base (NaOH) problem. The HCN is in the flask, and the NaOH serves as the titrant.

What is the pH of a solution of 25.0 mL of a 0.01 M solution of HCN after 50.0 mL of 0.05 M NaOH has been added?

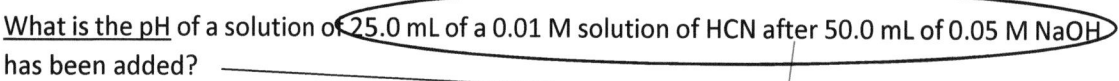

You know NaOH is the titrant because of the word "after" and "has been added"

Step 2. Use the stacking method to identify the quantity of each component.

Multiply the volume amount of each component by their concentration to yield the mmole values of each:

| Flask | HCN | 25.0 mL | × | 0.10 M | = | 2.5 mmol HCN |
| Titrant | NaOH | 50.0 mL | × | 0.05 M | = | 2.5 mmol NaOH |

Again, you can approach this next step in two ways: write the neutralization reaction with initial, changing, and resulting values, or simply subtract.

Step 2a. Use the neutralization reaction.

In the neutralization reaction, compare the quantities of HCN and NaOH. In this problem, the values are equal and both reactants are completed depleted. The only substance remaining is the conjugate base, CN^{-1}.

$$HCN\ (aq) + NaOH\ (aq) \rightarrow H_2O\ (l) + CN^-\ (aq)$$

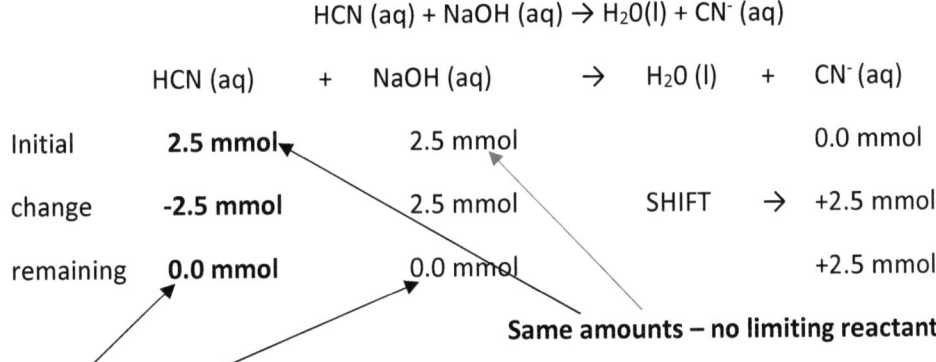

	HCN (aq)	+	NaOH (aq)	→	H_2O (l)	+	CN^- (aq)
Initial	2.5 mmol		2.5 mmol				0.0 mmol
change	-2.5 mmol		2.5 mmol		SHIFT →		+2.5 mmol
remaining	0.0 mmol		0.0 mmol				+2.5 mmol

Same amounts – no limiting reactant

No HCN nor NaOH remain in solution. Only component is the formation of the conjugate base CN^-.

Step 2b. Simply subtract.

Using the first table:

| Flask | HCN | 25.0 mL | × | 0.10 M | = | 2.5 mmol HCN |
| Titrant | NaOH | 50.0 mL | × | 0.05 M | = | 2.5 mmol NaOH |

Amount of CN^- = Amount Base Added

Subtract the values in the last column, and assess what remains. Thus,

2.5 mmol HCN

-2.5 mmol NaOH

0.0 mmol of either HCN or NaOH remain.

However, in weak/strong titrations, the amount of base added is equal to the amount of conjugate formed.

Step 4. Calculate CN⁻ concentration: TOTAL VOLUME

The total volume at the equivalence point is:

Total Volume = HCN mL + added NaOH mL = 25.0 + 50.0 = 75.0 mL

$$\text{Concentration of CN}^- = \frac{2.5 \text{ mmol}}{75.0 \text{ mL}} = 0.033$$

Step 5. Calculate the pH of the resulting flask solution.

The strong conjugate base, CN⁻, is the only component remaining at the equivalence point. To calculate the pH of the resulting solution, you must consider the hydrolysis reaction of CN⁻.

$$CN^- \text{ (aq)} + H_2O \text{ (l)} \rightleftharpoons HCN \text{ (aq)} + OH^- \text{ (aq)}$$

$$K_b = \frac{[HCN][OH^{-1}]}{[CN^{-1}]}$$

	CN⁻ (aq)	+ H₂O (l) ⇌	OH⁻ (aq) +	HCN (aq)
Initial				
Change				
Equilibrium				

Complete the ICE table for the shift, the change, x, and the equilibrium conditions.

	CN⁻ (aq)	+ H₂O (l) ⇌	OH⁻ (aq) +	HCN (aq)
Initial	0.033 M		0	0
	SHIFTS		→	
Change	-x		+x	+x
Equilibrium	0.033 - x		x	x

Once you have completed the table, you can now solve the problem for x.

Step 6. Place the equilibrium values from the ICE table into the equilibrium expression, and solve for x.

$$K_b = \frac{[HCN][OH^{-1}]}{[CN^{-1}]}$$

At this point, you do not know the K_b, but you do know K_a. Recall $K_a \times K_b = K_w$, and this must be utilized to prove the K_b value for the CN⁻ ion:

$$\frac{K_w}{K_a} = K_b = \frac{[HCN][OH^{-1}]}{[CN^{-1}]}$$

$$\frac{K_w}{K_a} = K_b = \frac{[HCN][OH^{-1}]}{[CN^{-1}]}$$

$$K_b = \frac{1.0 \times 10^{-14}}{4.9 \times 10^{-10}} = \frac{[HCN][OH^{-1}]}{[CN^{-1}]}$$

$$2.0 \times 10^{-5} = \frac{[x][x]}{[0.033-x]}$$

Upon first inspection of this equation, a polynomial exists, and you can absolutely use the quadratic equation to solve for x. However, you can check the value of x to see if it is negligible.

Step 7. Assess x negligibility.

Use the following equation to see if x can be deemed negligible:

$$\frac{Initial\ Concentration\ of\ Weak\ Base}{K_b\ of\ weak\ base} > 100$$

For this problem:

$$\frac{0.033\ M}{2.0 \times 10^{-5}} > 100$$

$$165 > 100$$

165 is greater than 100; thus, x is negligible and can be excluded from the equilibrium expression.

Step 6. Solve for x, and calculate pOH.

For weak base equilibria, the x in the expression stands for the [OH⁻]. Thus, you can determine the pOH and then the pH.

$$2.0 \times 10^{-5} = \frac{(x)^2}{(0.033-x)}$$

$$2.0 \times 10^{-5} = \frac{(x)^2}{(0.033)}$$

$$6.6 \times 10^{-7} = x^2$$

$$\sqrt{6.6 \times 10^{-7}} = \sqrt{x^2}$$

$$8.1 \times 10^{-4} = x$$

In this problem, x = [OH⁻] (strong conjugate base hydrolysis); thus, the pOH is:

$$pOH = -\log[OH^-]$$

$$pOH = -\log[8.1 \times 10^{-4}]$$

$$pOH = 3.09$$

Remember:
$$pH + pOH = 14$$

$$pH + 3.09 = 14$$

$$pH = 10.91$$

Section 7.2.4 After the Equivalence Point – Weak Acid/Strong Base

At this point in the titration, excess titrant or NaOH remains.

Sample Problem 7.8 pH after the equivalence point – weak acid/strong base

What is the pH of a solution of 25.0 mL of a 0.10 M solution of HCN after 80.0 mL of 0.05 M NaOH has been added? K_a of HCN = 4.9 × 10⁻¹⁰.

Step 1. Identify the type of acid/base titration, the flask solution and the titrant.

This is a weak acid (HCN)/strong base (NaOH) problem. The HCN is in the flask, and the NaOH serves as the titrant.

<u>What is the pH</u> of a solution of ⟨25.0 mL of a 0.01 M solution of HCN after 80.0 mL of 0.05M NaOH⟩ has been added? K_a of HCN = 4.9 × 10⁻¹⁰.

You know NaOH is the titrant because of the word "after" and "has been added"

Chapter 7. Titrations | 153

Step 2. Use the stacking method to identify the quantity of each component.

Multiply the volume amount of each component by their concentration to yield the mmole values of each:

Flask   HCN   25.0 mL   ×   0.10 M   =   2.5 mmol HCN

Titrant   NaOH   80.0 mL   ×   0.05 M   =   4.0 mmol NaOH

Again, you can approach this next step in two ways, write the neutralization reaction with initial, changing, and resulting values or simply subtract.

Step 2a. Use the neutralization reaction.

In the neutralization reaction, compare the quantities of HCN and NaOH. In this problem, the HCN now limits, and NaOH is in excess with a small amount of CN^- being formed.

$$HCN\ (aq) + NaOH\ (aq) \rightleftarrows H_2O\ (l) + CN^-\ (aq)$$

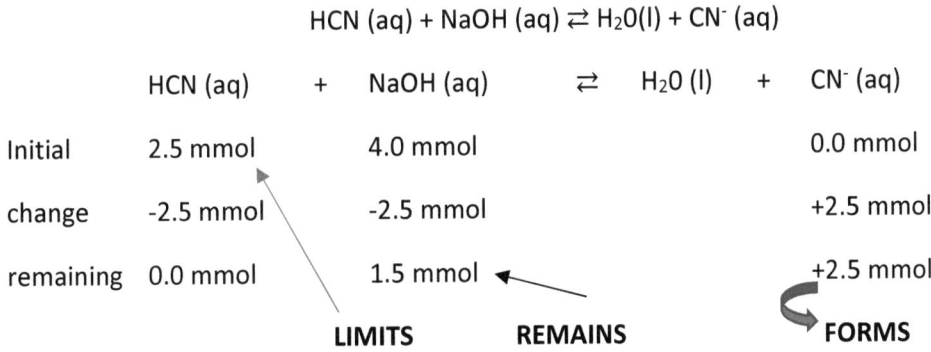

Step 2b. Simply subtract.

Using the first table:

Flask   HCN   25.0 mL   ×   0.10 M   =   2.5 mmol HCN

Titrant   NaOH   80.0 mL   ×   0.05 M   =   4.0 mmol NaOH

Subtract the values in the last column and assess what remains. Thus,

2.5 mmol HCN

-4.0 mmol NaOH

1.5 mmol NaOH remain

Note: Whichever component limits is the amount of conjugate formed.

Step 3. Calculate the TOTAL VOLUME in the flask.

From the table:

Flask	HCN	25.0 mL	x	0.10 M	=	2.5 mmol HCN
Titrant	NaOH	80.0 mL	x	0.05 M	=	4.0 mmol NaOH

Total Volume = HCN mL + added NaOH mL = 25.0 + 80.0 = 105.0 mL

Step 4. Calculate the pH of the resulting flask solution.

After the equivalence point in these weak acid/strong base titrations, both strong base [OH⁻] and conjugate base [CN⁻] are present, and you can consider both to contribute to the pH of the solution. However, the [OH⁻] so overwhelms the amount of CN⁻ that its contribution is negligible compared to the pH effect of the OH⁻.

Since the NaOH remains, solve for the pOH initially and then pH:

$$pOH = -\log[OH^-]$$

$$pH + pOH = 14$$

The CONCENTRATION is required, and it is the ratio of mmol of NaOH in Step 2 (1.5 mmol) to the total volume in Step 3 (105.0 mL). The OH⁻ concentration is the resulting concentration of the NaOH, which is:

$$\frac{mmol\ NaOH}{Total\ Volume\ in\ mL}$$

$$\frac{1.5\ mmol\ NaOH}{105.0\ mL} = 0.01429\ M\ NaOH$$

$$pOH = -\log(0.01429)$$

$$pOH = 1.85$$

$$pH = 14 - pOH$$

$$pH = 14 - 1.85 = 12.15$$

Section 7.3 Weak Base/Strong Acid Titrations

In these types of titrations, the weak base is in the flask, and the strong acid serves as the titrant. The equations you need to solve these problems are:

Starting Point: Weak Base Hydrolysis: $K_b = \dfrac{x^2}{[weak\ base]_{initial}}$; $x = [OH^-]$

Before Equivalence: Buffer: $pOH = pK_b + \log\dfrac{[conjugate\ acid]}{[weak\ base]}$

At Equivalence Point: Conjugate Acid Hydrolysis: $\dfrac{K_w}{K_b} = \dfrac{x^2}{[conjugate\ acid]}$; $x = [H_3O^+]$

After Equivalence Point: Strong Acid in Excess: $pH = -\log(H_3O^+)$

Section 7.3.1 Initial State – No Titrant – Weak Base/Strong Acid

Sample Problem 7.9 pH at the initial state – no titrant – weak base/strong acid

What is the pH of 25.0 mL of a 0.20 M solution of Methylamine, CH_3NH_2, before the addition of the titrant 0.15 M HCl? K_b for $CH_3NH_2 = 4.4 \times 10^{-4}$.

Step 1. Identify the type of acid/base titration, the flask solution, and the titrant.

In these problems, an acid/base or neutralization reaction occurs. The key is determining which reactant is the limiting reagent. For this problem, CH_3NH_2 and HCl are reactants and are a WEAK Base (K_b is given) and STRONG acid. Thus, solution will follow this row from the chart:

Type of Titration	Starting Point (no titrant)	Before Equivalence Point	Equivalence Point	After Equivalence Point
Weak Base Strong Acid	$K_b = x^2/[weak\ base]_i$ $x = [OH-]$ $pOH = -\log[OH^-]$ $pH = 14 - pOH$	$pOH = pK_b + \log(\dfrac{[conjugate\ acid]}{[weak\ base]})$	$K_w/K_b = x^2/[conj\ acid]$ $X = [H_3O^+]$ $pH = -\log[H_3O^+]$	$pH = -\log[H_3O^+]$

The methylamine, CH_3NH_2, is the flask solution, and as the problem states, the HCl serves as the titrant.

Step 2. Use the stacking method to identify the quantity of each component.

Multiply the volume amount of each component by its concentration to yield the mmole values of each:

Flask	CH_3NH_2	25.0 mL	×	0.20 M	=	5.0 mmol CH_3NH_2
Titrant	HCl	0.0 mL	×	0.15 M	=	0.0 mmol HCl

Since there is no HCl in the flask, the only component contributing to the pH is the CH_3NH_2.

Step 3. Solve for the pH of the solution in the flask.

There is solely 25.0 mL of 0.20 M CH_3NH_2 in the flask, which is a weak base. To get the pH of this solution, you must determine the equilibrium hydrolysis of CH_3NH_2:

$$CH_3NH_2\ (aq) + H_2O\ (l) \rightleftarrows CH_3NH_3^+\ (aq) + OH^-\ (aq)$$

Using an ICE table:

$$K_b = \frac{[CH_3NH_3^{+1}][OH^{-1}]}{[CH_3NH_2]}$$

	CH_3NH_2 (aq)	+ H_2O (l)	$\rightleftarrows$	$CH_3NH_3^+$(aq) +	OH^{-1}(aq)
Initial					
Change					
Equilibrium					

Write the initial conditions, the change and the equilibrium condition.

	CH_3NH_2 (aq)	+ H_2O (l)	⇌	$CH_3NH_3^+$ (aq) +	OH^{-1} (aq)
Initial	0.20 M			0	0
SHIFT			→		
Change	-x			+x	+x
Equilibrium	0.20 - x			+x	+x

Place the equilibrium values from the ICE table into the equilibrium expression, and solve for x.

$$K_b = \frac{[CH_3NH_3^{+1}][OH^{-1}]}{[CH_3NH_2]}$$

$$4.4 \times 10^{-4} = \frac{(x)(x)}{(0.20-x)}$$

Upon first inspection of this equation, a polynomial exists, and you can absolutely solve for these problems using the quadratic equation. Like in the prior examples, you can calculate the negligibility of x to determine if it can be excluded from both the equilibrium expression.

Step 4. Assess x negligibility.

Use the following equation to see if x can be deemed negligible:

$$\frac{\text{Initial Concentration of Weak Base}}{K_b \text{ of weak base}} > 100$$

For this problem:

$$\frac{0.20 \text{ M}}{4.4 \times 10^{-4}} > 100$$

$$455 > 100$$

455 is greater than 100; thus, x is negligible and can be excluded from the equilibrium expression.

Step 5. Solve for x, and calculate pH.

$$4.4 \times 10^{-4} = \frac{x^2}{(0.20)}$$

$$x^2 = 8.8 \times 10^{-5}$$

$$\sqrt{x^2} = \sqrt{8.8 \times 10^{-5}}$$

$$x = 9.4 \times 10^{-3}$$

In this problem, x = [OH⁻¹] (weak base hydrolysis); thus, the pOH is:

$$pOH = -\log[OH^-]$$

$$pOH = -\log[9.4 \times 10^{-3}]$$

$$pOH = 2.03$$

$$pH = 14 - 2.03 = 11.97$$

Section 7.3.2 Before the Equivalence Point – Weak Base/Strong Acid

NOTE: In a weak base/strong acid titration, the acid limits the reaction before the equivalence point. In this case, both weak base and its conjugate acid exist in the solution, which means a buffer forms.

$$CH_3NH_2 \text{ (aq)} + HCl \text{ (aq)} \rightleftharpoons CH_3NH_3^+ \text{ (aq)} + \cancel{Cl^- \text{(aq)}}$$

      remains   limits   forms

You can exclude the Cl⁻ from the equilibrium, because it serves as a weak conjugate that will not hydrolyze and will not affect the pH.

Sample Problem 7.10 pH before the equivalence point – weak acid/strong base

What is the pH of 25.0 mL of a 0.20 M solution of methylamine, CH_3NH_2, after 12.0 ml of 0.15 M HCl have been added? K_b for $CH_3NH_2 = 4.4 \times 10^{-4}$.

Step 1. Identify the type of acid/base titration, the flask solution, and the titrant.

This is a weak base (CH_3NH_2)/strong acid (HCl) problem. The CH_3NH_2 is in the flask, and the HCl serves as the titrant.

What is the pH of 25.0 mL of a 0.20 M solution of methylamine, CH_3NH_2, after 12.0 ml of 0.15 M HCl has been added? K_b for $CH_3NH_2 = 4.4 \times 10^{-4}$.

      You know HCl is the titrant because of the word "after" and "has been added"

Step 2. Use the stacking method to identify the quantity of each component.

Multiply the volume amount of each component by their concentration to yield the mmole values of each:

| Flask | CH_3NH_2 | 25.0 mL | × | 0.20 M | = | 5.0 mmol CH_3NH_2 |
| Titrant | HCl | 12.0 mL | × | 0.15 M | = | 1.8 mmol HCl |

Step 2a. Use the neutralization reaction.

In the neutralization reaction, compare the quantities of CH_3NH_2 and HCl. In this problem, the HCl is smaller and, thus, limits.

$$CH_3NH_2\ (aq) + HCl\ (aq) \rightleftharpoons CH_3NH_3^+\ (aq) + Cl^-\ (aq)$$

	CH_3NH_2 (aq)	+ HCl (aq)	→	$CH_3NH_3^+$ (aq)
Initial	5.0 mmol	1.8 mmol		0.0 mmol
change	-1.8 mmol	-1.8 mmol	SHIFTS →	+ 1.8 mmol
remaining	3.2 mmol	0.0 mmol		+1.8 mmol
	REMAINS	LIMITS		FORMS

Or Step 2b. Use the subtraction method.

Using this table:

| Flask | CH_3NH_2 | 25.0 mL | × | 0.20 M | = | 5.0 mmol CH_3NH_2 |
| Titrant | HCl | 12.0 mL | × | 0.15 M | = | 1.8 mmol HCl |

Again, looking solely at the last column, the amount of CH_3NH_2 remaining, and you can simply determine the amount of $CH_3NH_3^+$ formed:

5.0 mmol CH_3NH_2

1.8 mmol HCl ← $CH_3NH_3^+$ produced

Subtract values yield the amount of CH_3NH_2 remaining

5.0 mmol CH_3NH_2

-1.8 mmol HCl

3.2 mmol CH_3NH_2 - remaining

Since CH_3NH_2 and its conjugate $CH_3NH_3^+$ coexist in solution, this is a buffer, and you can use the Henderson-Hasselbalch equation to solve for pOH.

Step 3. Calculate the TOTAL VOLUME in the flask.

This step is not needed for buffer solutions, because the weak base and its conjugate are in the same volume. The mmole values of each will be ratioed to provide the given result.

Step 4. Calculate the pH of the resulting buffer solution in the flask.

Use the Henderson-Hasselbalch equation to solve for the pH:

$$pOH = pK_b + \log \frac{[conjugate\ acid]_{formed}}{[weak\ base]_{remaining}}$$

$$pOH = -\log(4.4 \times 10^{-4}) + \log \frac{(1.8\ mmol)}{(3.2\ mmol)}$$

$$pOH = 3.36 + (-0.2498)$$

$$pOH = 3.11$$

$$pH + pOH = 14$$

$$pH = 14 - pOH$$

$$pH = 14 - 3.11 = 10.89$$

Section 7.3.3 At the Equivalence Point – Weak Base/Strong Acid

The equivalence point of a titration is where the mmoles of the flask solution equal the mmoles of the titrant. The set of equations that you will use to solve at this point in the titration are:

$$\frac{K_w}{K_b} = \frac{x^2}{[conjugate\ acid]}$$

$$x = [H_3O^+]$$

$$pH = -\log[H_3O^+]$$

Sample Problem 7.11 pH at the equivalence point – weak base/strong acid

What is the pH of 25.0 mL of a 0.20 M solution of methylamine, CH_3NH_2, after the addition of 33.33 ml of 0.15 M HCl? K_b for $CH_3NH_2 = 4.4 \times 10^{-4}$.

Chapter 7. Titrations | **161**

Step 1. Identify the type of acid/base titration, the flask solution, and the titrant.

This is a weak base (CH_3NH_2)/strong acid (HCl) problem. The CH_3NH_2 is in the flask and the HCl serves as the titrant.

What is the pH of 25.0 mL of a 0.20 M solution of methylamine, CH_3NH_2, after the addition of 33.33 ml of 0.15 M HCl? K_b for $CH_3NH_2 = 4.4 \times 10^{-4}$?

You know HCl is the titrant because of the word "after" and "addition"

Step 2. Use the stacking method to identify the quantity of each component.

Multiply the volume amount of each component by its concentration to yield the mmole values of each:

Flask CH_3NH_2 25.00 mL   x   0.20 M   =   5.0 mmol CH_3NH_2

Titrant   HCl   33.33 mL   x   0.15 M   =   5.0 mmol HCl

Again, you can approach this next step in two ways: write the neutralization reaction with initial, changing, and resulting values, or simply subtract.

Step 2a. Use the neutralization reaction.

In the neutralization reaction, compare the quantities of CH_3NH_2 and HCl. In this problem, the values are equal and both reactants are completely depleted. Only $CH_3NH_3^+$ remains.

$$CH_3NH_2\ (aq) + HCl\ (aq) \rightleftarrows CH_3NH_3^+\ (aq)$$

	CH_3NH_2 (aq)	+	HCl (aq)	→	$CH_3NH_3^+$ (aq)
Initial	5.0 mmol		5.0 mmol		0.0 mmol
change	-5.0 mmol		-5.0 mmol	SHIFTS →	+5.0 mmol
remaining	0.0 mmol		0.0 mmol		5.0 mmol

Same amounts – no limiting reactant

No CH_3NH_2 nor HCl remain in solution. The only component is the formation of the conjugate acid $CH_3NH_3^+$.

Step 2b. Simply subtract.

Using the first table:

Flask	CH_3NH_2	25.00 mL	x	0.20 M	=	5.0 mmol CH_3NH_2
Titrant	HCl	33.33 mL	x	0.15 M	=	5.0 mmol HCl

Amount of $CH_3NH_3^+$ = Amount Acid Added

Subtract the values in the last column, and assess what remains. Thus,

5.0 mmol CH_3NH_2

-5.0 mmol HCl

0.0 mmol of either CH_3NH_2 or HCl remain

In weak base/strong acid titrations, the amount of acid added equals the amount of conjugate formed.

Step 4. Calculate $CH_3NH_3^+$ concentration: TOTAL VOLUME.

The total volume at the equivalence point is:

Total Volume = CH_3NH_2 mL + added HCl mL = 25.0 + 33.33 = 58.33 mL

Concentration of $CH_3NH_3^+ = \dfrac{5.0 \; mmol}{58.33 \; mL} = 0.0857$ M

Step 5. Calculate the pH of the resulting flask solution.

The strong conjugate acid, $CH_3NH_3^+$, is the only component remaining at the equivalence point. To calculate the pH of the resulting solution, you must consider the hydrolysis reaction of $CH_3NH_3^+$.

$$CH_3NH_3^+ \; (aq) + H_2O \; (l) \rightleftarrows CH_3NH_2 \; (aq) + H_3O^+ \; (aq)$$

$$K_a = \frac{[CH_3NH_2][H_3O^{+1}]}{[CH_3NH_3^{+1}]}$$

$$CH_3NH_3^+ \text{ (aq)} + H_2O \text{ (l)} \rightleftharpoons CH_3NH_2 \text{ (aq)} + H_3O^+ \text{ (aq)}$$

Initial

Change

Equilibrium

Complete the ICE table for the shift, the change, x, and the equilibrium conditions.

$$CH_3NH_3^+ \text{ (aq)} + H_2O \text{ (l)} \rightleftharpoons CH_3NH_2 \text{ (aq)} + H_3O^+ \text{ (aq)}$$

	$CH_3NH_3^+$		CH_3NH_2	H_3O^+
Initial	0.0857 M		0	0
	SHIFTS	→		
Change	-x		+x	+x
Equilibrium	0.0857 - x		x	x

Once you have completed the table, you can now solve the problem for x.

Step 6. Place the equilibrium values from the ICE table into the equilibrium expression, and solve for x.

$$K_a = \frac{[CH_3NH_2][H_3O^{+1}]}{[CH_3NH_3^{+1}]}$$

At this point, you do not know the K_b, but you do know K_a. Recall $K_a \times K_b = K_w$, and you must utilize this to prove the K_b value for the CN⁻ ion.

$$\frac{K_w}{K_b} = K_a = \frac{[CH_3NH_2][H_3O^{+1}]}{[CH_3NH_3^{+1}]}$$

$$K_a = \frac{1.0 \times 10^{-14}}{4.4 \times 10^{-4}} = \frac{[CH_3NH_2][H_3O^{+1}]}{[CH_3NH_3^{+1}]}$$

$$K_a = \frac{1.0 \times 10^{-14}}{4.4 \times 10^{-4}} = \frac{(x)(x)}{[0.0857-x]}$$

$$2.27 \times 10^{-11} = \frac{x^2}{[0.0857-x]}$$

Upon first inspection of this equation a polynomial exists, and you can absolutely use the quadratic equation to solve for x. However, you can check the value of x to see if it is negligible.

Step 7. Assess x negligibility.

Use the following equation to see if x can be deemed negligible:

$$\frac{\text{Initial Concentration of conjugate acid}}{K_a \text{ of conjugate acid}} > 100$$

For this problem:

$$\frac{0.0857 \text{ M}}{2.27 \times 10^{-11}} > 100$$

$$3.77 \times 10^9 > 100$$

3.77×10^9 is greater than 100; thus, x is negligible and you can exclude it from the equilibrium expression.

Step 8. Solve for x, and calculate pH.

For the strong conjugate acid equilibria, the x in the expression stands for the $[H_3O^+]$, and you can directly determine the pH.

$$2.27 \times 10^{-11} = \frac{(x)^2}{(0.0857 - \cancel{x})}$$

$$2.27 \times 10^{-11} = \frac{(x)^2}{(0.0857)}$$

$$1.95 \times 10^{-12} = x^2$$

$$\sqrt{1.95 \times 10^{-12}} = \sqrt{x^2}$$

$$1.40 \times 10^{-6} = x$$

In this problem, $x = [H_3O^+]$ (strong conjugate acid hydrolysis); thus, the pH is:

$$pH = -\log[H_3O^+]$$

$$pH = -\log[1.40 \times 10^{-6}]$$

$$pH = 5.86$$

Section 7.3.4 After the Equivalence Point – Weak Base/Strong Acid

At this point in the titration, excess titrant or HCl remains.

Sample Problem 7.12 pH after the Equivalence Point – Weak Base/Strong Acid

What is the pH of 25.0 mL of a 0.20 M solution of methylamine, CH_3NH_2, after the addition of 55.0 ml of 0.15 M HCl? K_b for CH_3NH_2 = 4.4 x 10^{-4}.

Step 1. Identify the type of acid/base titration, the flask solution, and the titrant.

This is a weak base (CH_3NH_2)/strong acid (HCl) problem. The CH_3NH_2 is in the flask, and the HCl serves as the titrant

What is the pH of 25.0 mL of a 0.20 M solution of methylamine, CH_3NH_2, after the addition of 55.0 ml of 0.15 M HCl? K_b for CH_3NH_2 = 4.4 x 10^{-4}.

You know HCl is the titrant because of the word "after" and "addition"

Step 2. Use the stacking method to identify the quantity of each component.

Multiply the volume amount of each component by its concentration to yield the mmole values of each:

Flask	CH_3NH_2	25.0 mL	x	0.20 M	=	5.0 mmol CH_3NH_2
Titrant	HCl	55.0 mL	x	0.15 M	=	8.25 mmol HCl

Again, you can approach this next step in two ways: write the neutralization reaction with initial, changing, and resulting values, or simply subtract.

Step 2a. Use the neutralization reaction.

In the neutralization reaction, compare the quantities of CH_3NH_2 and HCl. In this problem, the CH_3NH_2 now limits, and HCl is in excess with a small amount of $CH_3NH_3^+$ being formed.

$$CH_3NH_2 \text{ (aq)} + HCl \text{ (aq)} \rightleftharpoons CH_3NH_3^+ \text{ (aq)}$$

	CH₃NH₂ (aq)	+ HCl (aq)	→	CH₃NH₃⁺ (aq)
Initial	5.0 mmol	8.25 mmol		0.0 mmol
change	-5.0 mmol	-5.0 mmol	SHIFTS →	+5.0 mmol
remaining	0.0 mmol	3.25 mmol		5.0 mmol
	LIMITS	REMAINS		FORMS

Step 2b. Simply subtract.

Using the first table:

| Flask | CH₃NH₂ | 25.0 mL | x | 0.20 M | = | 5.0 mmol CH₃NH₂ |
| Titrant | HCl | 55.0 mL | x | 0.15 M | = | 8.25 mmol HCl |

Subtract the values in the last column and assess what remains. Thus,

5.0 mmol CH₃NH₂

-8.25 mmol HCl

3.25 mmol HCl remain

Note: Whichever component limits is the amount of conjugate formed.

Step 3. Calculate the TOTAL VOLUME in the flask.

From the table:

| Flask | CH₃NH₂ | 25.0 mL | x | 0.20 M | = | 5.0 mmol CH₃NH₂ |
| Titrant | HCl | 55.0 mL | x | 0.15 M | = | 8.25 mmol HCl |

Total Volume = CH₃NH₂ mL + added HCl mL = 25.0 + 55.0 = 80.0 mL

Step 4. Calculate the pH of the resulting flask solution.

After the equivalence point in these weak base/strong acid titrations, both a strong acid [H_3O^+] and conjugate acid [$CH_3NH_3^+$] are present. And, you can consider both to contribute to the pH of the solution. However, the [H_3O^+] overwhelms the amount of [$CH_3NH_3^+$] such that its contribution proves negligible compared to the pH effect of the [H_3O^+].

Since the HCl remains, you solve for the pH with:

$$pH = -\log[H_3O^+]$$

The CONCENTRATION is required, and it is the ratio of mmol of HCl in Step 2 (3.25 mmol) to the total volume in Step 3 (80.0 mL). The **H_3O^+** concentration is the resulting concentration of the HCl, which is:

$$\frac{mmol\ HCl}{Total\ Volume\ in\ mL}$$

$$\frac{3.25\ mmol\ HCl}{80.0\ mL} = 0.0406\ M\ HCl$$

$$pH = -\log(0.0406) = 1.39$$

Section 7.4 Typical Problems and Student Problems

In the prior sections, the solution methodology was provided for three different types of titrations. However, when these types of problems are given in the textbook or on exams, the problems can come from any circumstance. So, the student must systematically determine first what type of titration and, then, where in the titration the values land the solution. You can use the following methodology to identify and solve these titration problems.

Sample Problem 7.13 Titration Problem

What is the pH of the solution after 30.0 ml of 0.25 M acetic acid, CH_3COOH, is titrated with 30.0 ml of 0.20 M NaOH? K_a of Acetic Acid = 1.8×10^{-5}.

Step 1. Identify the type of titration, the solution in the flask, and the titrant.

What is the pH of the solution after 30.0 ml of 0.25 M acetic acid, CH_3COOH, is titrated with 30.0 ml of 0.20 M NaOH? K_a of Acetic Acid = 1.8×10^{-5}.

The problem states acetic acid (K_a is given) and NaOH. Acetic acid is a weak acid (K_a), and NaOH is a strong base. **WEAK ACID/STRONG BASE**

NaOH is the titrant, and acetic acid is in the flask.

Step 2. Use the stacking method to identify the quantity of each component and where in the titration the problems occur.

Multiply the volume amount of each component by their concentration to yield the mmole values of each:

Flask	CH_3COOH	30.0 mL	x	0.25 M	=	7.5 mmol CH_3COOH
Titrant	NaOH	30.0 mL	x	0.20 M	=	6.0 mmol NaOH

Assessing the values, NaOH has the smaller amount of mmols and limits the maximum amount that can react; thus, the point in the titration that exists for this particular problem is BEFORE the EQUIVALENCE POINT: ergo a BUFFER solution exists.

Step 2a. Use the neutralization reaction.

In the neutralization reaction, compare the quantities of CH_3COOH and NaOH. In this problem, the NaOH is smaller and, thus, limits.

$$CH_3COOH\ (aq) + NaOH\ (aq) \rightleftharpoons H_2O\ (l) + CH_3COO^-\ (aq)$$

	CH₃COOH(aq)	+ NaOH (aq)	→	H₂O (l)	+	CH₃COO⁻ (aq)
Initial	7.5 mmol	6.0 mmol				0.0 mmols
change	-6.0 mmol	-6.0 mmol		SHIFT →		+6.0 mmoles
remaining	1.5 mmol	0.0 mmol				6.0 mmoles
	REMAINS	**LIMITS**				**FORMS**

Or Step 2b. Use the subtraction method.

Using this table:

| Flask | CH₃COOH | 30.0 mL | × | 0.25 M | = | 7.5 mmol CH₃COOH |
| Titrant | NaOH | 30.0 mL | × | 0.20 M | = | 6.0 mmol NaOH |

Again, looking solely at the last column, you can simply determine the amount of CH₃COOH remaining and the amount of CH₃COO⁻ formed:

7.5 mmol CH₃COOH

6.0 mmol NaOH

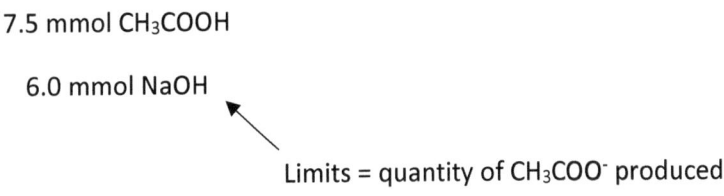

Limits = quantity of CH₃COO⁻ produced

Subtract values yields amount of CH₃COOH remaining.

7.5 mmol CH₃COOH

-6.00 mmol NaOH
_____

1.5 mmol CH₃COOH – remaining

Since CH₃COOH and its conjugate CH₃COO⁻ coexist in solution, this is a buffer, and the Henderson-Hasselbalch equation can be used to solve for pH.

Step 3. Calculate the TOTAL VOLUME in the flask.

This step is not needed for buffer solutions, because the weak acid and its conjugate are in the same volume. The mmole values of each will be ratioed to provide the given result.

Step 4. Calculate the pH of the resulting buffer solution in the flask.

The equation to solve for pH is the Henderson-Hasselbalch equation:

$$pH = pK_a + \log \frac{[conjugate\ base]_{formed}}{[weak\ acid]_{remaining}}$$

$$pH = -\log(1.8 \times 10^{-5}) + \log \frac{(6.0\ mmol)}{(1.5\ mmol)}$$

$$pH = 4.74 + 0.602$$

$$pH = 5.34$$

Sample Problem 7.14 Titration problem

What is the pH of the solution of 25.0 ml of 0.45 M HClO₃ to which 30.0 ml of 0.35 M NaOH has been added?

Step 1. Identify the type of titration, the solution in the flask, and the titrant.

What is the pH of the solution of 25.0 ml of 0.45 M HClO₃ to which 30.0 ml of 0.35 M NaOH has been added?

In the problem, NaOH has been added to a solution of HClO₃. NaOH is a strong base, HClO₃ is a strong acid—the problem gives no K_a or K_b values. Thus, this is a STRONG ACID/STRONG BASE titration.

Step 2. Use the stacking method to identify the quantity of each component and where in the titration the problem occurs.

Multiply the volume amount of each component by its concentration to yield the mmole values of each:

Flask	HClO₃	25.0 mL	x	0.45 M	=	11.25 mmol HClO₃
Titrant	NaOH	30.0 mL	x	0.35 M	=	10.50 mmol NaOH

Assessing the values, NaOH limits thus BEFORE EQUIVALENCE POINT

Step 2a. Using the neutralization reaction

In the neutralization reaction, compare the quantities of HClO₃ and NaOH. In this problem, the NaOH is smaller and, thus, limits.

$$HClO_3 \text{ (aq)} + NaOH \text{ (aq)} \rightleftarrows H_2O \text{ (l)} + NaClO_3 \text{ (aq)}$$

	$HClO_3$ (aq)	+	NaOH (aq)	→	H_2O (l)	+	$NaClO_3$ (aq)
Initial	11.25 mmol		10.50 mmol				
change	-10.50 mmol		-10.50 mmol				
remaining	0.75 mmol		0.0 mmol				
	REMAINS		**LIMITS**				

Or Step 2b. Use the subtraction method.

Using this table:

Flask	$HClO_3$	25.0 mL	x	0.45 M	=	11.25 mmol $HClO_3$
Titrant	NaOH	30.0 mL	x	0.35 M	=	10.50 mmol NaOH

Again, looking solely at the last column, you can simply determine the amount of $HClO_3$ remaining by subtracting the two values:

11.25 mmol $HClO_3$

10.50 mmol NaOH ← Limits

11.25 mmol $HClO_3$

<u>-10.50 mmol NaOH</u>

0.75 mmol $HClO_3$ remains

Step 3. Determine the TOTAL volume.

From the table:

Flask	$HClO_3$	25.0 mL	x	0.45 M	=	11.25 mmol $HClO_3$
Titrant	NaOH	30.0 mL	x	0.35 M	=	10.50 mmol NaOH

Total Volume = $HClO_3$ mL + added NaOH mL = 25.0 + 30.0 = 55.0 mL

Step 4. Calculate the HClO$_3$ concentration and pH of the resulting flask solution.

The equation to solve for pH is:

$$pH = -\log[H_3O^+]$$

Thus, the CONCENTRATION is required, and it is the ratio of mmol of HCl in Step 2 (0.75 mmol) to the total volume in Step 3 (55.0 mL). The H$_3$O$^+$ concentration is the resulting concentration of the HClO$_3$, which is:

$$\frac{mmol\ HClO_3}{Total\ Volume\ in\ mL}$$

$$\frac{0.75\ mmol\ HClO_3}{55.0\ mL} = 0.0136\ M$$

$$pH = -\log(0.0136)$$

$$pH = 1.87$$

STUDENT PROBLEMS

1. Calculate the pH of 20.0 ml of a 0.50 M solution of HF to which 50.0 ml of 0.05 M NaOH has been added. K$_a$ of HF = 6.8 x 10^{-4}.
2. A 30.0 ml sample of 0.25 M formic acid, HCOOH, is titrated with 0.30 M NaOH. What is the pH of the solution after 25.0 ml of the NaOH are added? K$_a$ formic acid = 1.7 x 10^{-4}.
3. What is the pH after 40.0 ml of 0.10 M NaOH is added to 30.0 ml of 0.20 M HI?
4. What is the pH of 50.0 ml of a 0.25 M NH$_3$ solution to which 35.0 ml of 0.45 M HBr has been added? K$_b$ of NH$_3$ = 1.8 x 10^{-5}.

(Answers: 1. 2.69; 2. 8.45; 3. 1.54; 4. 1.42)

Chapter 8. Solubility Equilibria

Section 8.1 Determination of K_{sp}

The final equilibrium expression covered in this semester of chemistry is solubility equilibria defined by the equilibrium constant K_{sp}. These equilibria help define the solubility of inorganic salts that have very limited solubility in water. These solutions follow the methodology that was covered in Chapter 4 of this problem solving manual.

Solving Tips:

1. Write equilibrium expressions, as learned in Chapter 4: Products/Reactants.
2. Exclude solids and liquids from the equilibrium expression.
3. Solubility = grams of salt/L of water.
4. Molar solubility = moles of salt/L of water.
5. K_{sp} values for many of these salts are very small values.
6. When solving for x in these types of problems, x is always the MOLAR SOLUBILITY.

Sample Problem 8.1 Determination of Molar Solubility from K_{sp}

What is the molar solubility of BaF_2? The K_{sp} of BaF_2 is 1.0×10^{-6}.

Step 1. Identify the type of equilibrium problem and the known and unknown variables.

When a problem provides either a K_{sp} value or asks for solubility or molar solubility, it requires a solubility equilibrium.

What is the molar solubility of BaF_2? The K_{sp} of BaF_2 is 1.0×10^{-6}.

Step 2. Write the solubility reaction and the equilibrium expression. Create an ICE table.

$$BaF_2 \, (s) \rightleftharpoons Ba^{+2} \, (aq) + 2F^- \, (aq)$$

$$K_{sp} = [Ba^{+2}][F^-]^2$$

	BaF_2 (s)	$\rightleftharpoons$	Ba^{+2} (aq) +	$2F^-$ (aq)
Initial				
Change				
Equilibrium				

Step 3. Fill in the initial conditions, the change, and the equilibrium values.

Initially, there is only BaF$_2$. Thus, the equilibrium SHIFTS RIGHT to form the ions in solution.

	BaF$_2$ (s)	⇌	Ba^{+2} (aq) +	2F$^-$ (aq)
Initial			0	0
			Shifts →	
Change				
Equilibrium				

Assess the change, x, and equilibrium condition.

Because the equilibrium shifts right, the BaF$_2$ disassociates into the ions Ba^{+2} and F$^-$. Since BaF$_2$ is a solid, it is excluded from the equilibrium expression, and the only change occurs for Ba^{+2} and F$^-$ which is positive. Remember to include the coefficient prior to each component in the change. E.g., 2F$^-$ means a gain of 2x.

	BaF$_2$ (s)	⇌	Ba^{+2} (aq) +	2F$^-$ (aq)
Initial			0	0
			SHIFT →	
Change			+ x	+2x
Equilibrium			x	2x

Once you have completed the table, you can now solve the problem for x.

Step 4. Place the equilibrium values from the ICE table into the equilibrium expression, and solve for x.

	$BaF_2\ (s)$ ⇌	$Ba^{+2}\ (aq)\ +$	$2F^-\ (aq)$
Initial		0	0
		SHIFT →	
Change		+ x	+2x
Equilibrium		x	2x

$$K_{sp} = [Ba^{+2}][F^-]^2$$

$$K_{sp} = [x][2x]^2$$

$$1.0 \times 10^{-6} = [x]4x^2$$

$$1.0 \times 10^{-6} = 4x^3$$

$$2.5 \times 10^{-7} = x^3$$

$$\sqrt[3]{2.5 \times 10^{-7}} = \sqrt[3]{x^3}$$

$$x = 6.30 \times 10^{-3}$$

When solving for x in these types of problems, x is always the MOLAR SOLUBILITY.

Sample Problem 8.2 Determination of the K_{sp} from Molar Solubility

If the molar solubility of $Fe(OH)_3$ is 7.38×10^{-5} mol/L what is its K_{sp} value?

Step 1. Identify the type of equilibrium problem and the known and unknown variables.

When a problem provides either a K_{sp} value or asks for solubility or molar solubility, it will be a solubility equilibrium.

$$Fe(OH)_3\ (s) ⇌ Fe^{+3}\ (aq) + 3OH^-\ (aq)$$

Step 2. Write the solubility reaction and the equilibrium expression. Create an ICE table.

$$Fe(OH)_3\ (s) ⇌ Fe^{+3}\ (aq) + 3OH^-\ (aq)$$

$$K_{sp} = [Fe^{+3}][OH^-]^3$$

Chapter 8. Solubility Equilibria | **177**

$$Fe(OH)_3\ (s) \rightleftarrows Fe^{+3}\ (aq) + 3OH^-\ (aq)$$

Initial

Change

Equilibrium

Step 3. Fill in the initial conditions, the change, and the equilibrium values.

Initially, there is only $Fe(OH)_3$. Thus, the equilibrium SHIFTS RIGHT to form the ions in solution.

	$Fe(OH)_3\ (s)$	$\rightleftarrows$	$Fe^{+3}\ (aq)\ +$	$3OH^-\ (aq)$
Initial	✕		0	0
		Shifts →		
Change	✕			
Equilibrium				

Assess the change, x, and equilibrium condition.

Because the equilibrium shifts right, the $Fe(OH)_3$ disassociates into the ions Fe^{+3} and OH^-. Since $Fe(OH)_3$ is a solid, it is excluded from the equilibrium expression, and the only change occurs for Fe^{+3} and OH^{-1} which is positive. Remember to include the coefficient prior to each component in the change (e.g., 3 OH^- means a gain of 3x).

	$Fe(OH)_3\ (s)$	$\rightleftarrows$	$Fe^{+3}\ (aq)\ +$	$3OH^-\ (aq)$
Initial	✕		0	0
		SHIFT →		
Change	✕		+ x	+3x
Equilibrium			x	3x

Once you have completed the table, you can now solve the problem for x.

Step 4. Place the equilibrium values from the ICE table into the equilibrium expression, define x, and solve for K_{sp}.

	$Fe(OH)_3$ (s)	⇌	Fe^{+3} (aq) +	$3OH^-$ (aq)
Initial			0	0
			SHIFT →	
Change			+ x	+3x
Equilibrium			x	3x

$$K_{sp} = [Fe^{+3}][\,3OH^-]^3$$

$$K_{sp} = [x][3x]^3$$

$$K_{sp} = [x]27x^3$$

$$K_{sp} = 27x^4$$

$$K_{sp} = 27(7.38 \times 10^{-5})^4$$

$$K_{sp} = 8.0 \times 10^{-16}$$

Sample Problem 8.3 Determination of Solubility from K_{sp}

What is the solubility of AgCl? K_{sp} of AgCl = 1.8×10^{-10}.

Step 1. Identify the type of equilibrium problem and the known and unknown variables.

When a problem provides either a K_{sp} value or asks for solubility or molar solubility, it will be a solubility equilibrium.

What is the solubility of AgCl? K_{sp} of AgCl = 1.8×10^{-10}.

Step 2. Write the solubility reaction and the equilibrium expression. Create an ICE table.

$$AgCl\ (s) \rightleftarrows Ag^{+1}\ (aq) + Cl^-\ (aq)$$

$$K_{sp} = [Ag^{+1}][\,Cl^-]$$

$$AgCl\ (s) \rightleftarrows Ag^{+1}\ (aq) + Cl^-\ (aq)$$

Initial

Change

Equilibrium

Step 3. Fill in the initial conditions, the change, and the equilibrium values.

Initially, there is only AgCl. Thus, the equilibrium SHIFTS RIGHT to form the ions in solution.

	$AgCl\ (s)$ $\rightleftarrows$	$Ag^{+1}\ (aq)\ +$	$Cl^-\ (aq)$
Initial		0	0
		Shifts →	
Change		+x	+x
Equilibrium		+x	+x

Assess the change, x, and equilibrium condition.

Because the equilibrium shifts right, the AgCl disassociates into the ions Ag^{+1} and Cl^-. Since AgCl is a solid, it is excluded from the equilibrium expression, and the only change occurs for Ag^{+1} and Cl^- which is positive.

	$AgCl\ (s)$ $\rightleftarrows$	$Ag^{+1}\ (aq)\ +$	$Cl^-\ (aq)$
Initial		0	0
		SHIFT →	
Change		+ x	+x
Equilibrium		x	x

Once you have completed the table, you can now solve the problem for x.

Step 4. Place the equilibrium values from the ICE table into the equilibrium expression, define x, and solve for K_{sp}.

	AgCl (s) ⇌	Ag^{+1} (aq) +	Cl$^-$ (aq)
Initial		0	0
Change		+ x	+x
Equilibrium		x	x

SHIFT →

$$K_{sp} = [Ag^{+1}][Cl^-]$$

$$K_{sp} = [x][x]$$

$$1.8 \times 10^{-10} = x^2$$

$$\sqrt{1.8 \times 10^{-10}} = \sqrt{x^2}$$

$$x = 1.34 \times 10^{-5} \text{ (molar solubility)}$$

Solubility = Molar Solubility x Molar Mass of salt

$$\text{Solubility} = (1.34 \times 10^{-5} \tfrac{mole}{L})(143.32 \tfrac{grams}{mole})$$

Solubility = 1.92×10^{-3} grams/L

STUDENT PROBLEMS

1. What is the molar solubility of copper (II) hydroxide? $K_{sp} = 2.6 \times 10^{-19}$
2. What is the solubility of Magnesium oxalate? $K_{sp} = 8.5 \times 10^{-5}$
3. The molar solubility for aluminum hydroxide is 3.61×10^{-9}; what is the solubility product?
(Answers: 1. 4.02×10^{-7} M; 2. 1.04 grams/L; 3. 4.59×10^{-33})

Section 8.2 Precipitation Problems

In the first semester of chemistry, the solubility of salts was a conceptual problem in which a chart was used to determine if a salt was soluble or insoluble. This chart is short sited, because every salt has some solubility in water, some very little and some quite large. In this section, problems will be presented to calculate the value at which a salt becomes insoluble or precipitates from solution. Remember, a low-soluble salt with a small K_{sp} will precipitate before a salt with higher solubility or large K_{sp}. In these problems, calculate the reaction quotient, Q, for the known concentrations, and then compare it to the known K_{sp} value for the sparingly soluble salt.

Solving Tips

If Q is greater than K_{sp}, a precipitate will form.

If Q is equal to or less than K_{sp}, no precipitate forms.

Sample Problem 8.4 Precipitation prediction

Will a precipitate form when 150.0 mL of 3.3×10^{-3} M $Pb(NO_3)_2$ are added to 150.0 mL of 4.44×10^{-4} M K_2S? K_{sp} of PbS = 2.5×10^{-27}, and KNO_3 is a highly soluble salt.

Step 1. Identify the type of equilibrium problem and the known and unknown variables.

The first part of this problem asked if a precipitate will form; thus, this is a solubility and K_{sp}-type problem. The concentration of all components is known; if a precipitate or a solid form is unknown. This problem requires to look at the reaction quotient, Q, versus the known K_{sp} value.

Step 2. Write the precipitation reaction and identify the least soluble salt.

Precipitation reaction:

$$Pb(NO_3)_2 \text{ (aq)} + K_2S \text{ (aq)} \rightarrow PbS \text{ (s)} + KNO_3 \text{ (aq)}$$

By using the solubility table, you can determine PbS to be the more insoluble salt and, thus, acquires the (s) designation.

Step 3. Calculate Q for the sparingly soluble salt.

The K_{sp} and equilibrium expression for PbS is:

$$PbS\ (s) \rightleftharpoons Pb^{+2}\ (aq) + S^{-2}\ (aq)$$

$$K_{sp} = [Pb^{+2}] \times [S^{-2}] = Q$$

There is no need for an ICE table in these problems, because only the concentration of each ion is important to solve the problem.

Step 3a. Determine the concentration of the Pb^{+2} and S^{-2}.

Concentration of Pb^{+2}:

$$[Pb^{+2}] = \frac{ml_{initial} \times Concentration_{initial}}{Total\ Volume\ after\ Mixing}$$

$$[Pb^{+2}] = \frac{(150.0\ ml)(3.33 \times 10^{-3})}{150.0\ ml + 150.0\ ml}$$

$$[Pb^{+2}] = \frac{0.4995}{300.0\ ml}$$

$$[Pb^{+2}] = 1.66 \times 10^{-3}\ M$$

Concentration of S^{-2}:

$$[S^{-2}] = \frac{ml_{initial} \times Concentration_{initial}}{Total\ Volume\ after\ Mixing}$$

$$[S^{-2}] = \frac{(150.0\ ml)(4.44 \times 10^{-4})}{150.0\ ml + 150.0\ ml}$$

$$[S^{-2}] = \frac{0.0666}{300.0\ ml}$$

$$[S^{-2}] = 2.22 \times 10^{-4}\ M$$

Step 4. Calculate Q, and assess versus K_{sp}.

$$Q = [Pb^{+2}] \times [S^{-2}]$$

$$Q = (1.66 \times 10^{-3})(2.22 \times 10^{-4})$$

$$Q = 3.70 \times 10^{-7}$$

Step 5. Q versus K_{sp}.

$$Q = 3.70 \times 10^{-7}$$

$$K_{sp}\ of\ PbS = 2.5 \times 10^{-27}$$

$$Q\ (3.70 \times 10^{-7}) > K_{sp}\ (2.5 \times 10^{-27})$$

A precipitate will form.

STUDENT PROBLEMS

1. Will a precipitate form when 500.0 mL of 5.0×10^{-5} M NiNO$_3$ are added to 400.0 mL of 2.5×10^{-7} M Na$_2$S? K_{sp} of NiS = 3.0×10^{-19}
2. Will a precipitate form 800.0 mL of 1.1×10^{-3} M BaCl$_2$ are added to 700.0 mL of 1.2×10^{-3} M Na$_2$SO$_4$? K_{sp} of BaSO$_4$ = 1.1×10^{-10}
3. Calculate the minimum concentration of Pb^{+2} that must be added to 0.105 M NaCl in order to precipitate PbCl$_2$. K_{sp} of PbCl$_2$ = 1.6×10^{-5}
   (Answers: 1. Yes; 2. Yes; 3. 1.45×10^{-3} M)

Chapter 9. Thermodynamics II

Section 9.1 Entropy and the Second Law of Thermodynamics

In the First Law of Thermodynamics, problems were present with regards to internal energy (U) and enthalpy (H). In this chapter of problems, the second law of thermodynamics is covered, which focuses primarily upon entropy (S).

Solving Tips

1. The main goal in each of these problems is to assess the sign of S. When positive, the reaction will be spontaneous in the direction written. If negative, the reaction will be non-spontaneous in the direction written.
2. You can consider entropy in two ways; the more energetic a system, the more entropy it possesses. From a statistical approach, the greater the number of configurations or possible states, the greater the entropy for the system.
3. Entropy is a positive quantity for all elements and compounds. It is never a negative quantity.
4. Entropy can be zero only if absolute zero (0 K) is achieved (third law of thermodynamics).
5. When working on problems with reactions, the change in the entropy of the overall reaction can be negative or positive, thus, inferring the spontaneity of that reaction as written.
6. The temperature of a reaction is a prime indicator to assess spontaneity. The phase change of liquid water to ice at 25°C is a non-spontaneous reaction. However, this same phase change at -1.0°C is spontaneous.
7. The units for entropy, S, are $\frac{J}{K\,mole}$, while Enthalpy, H, and Gibbs free energy, G, are $\frac{kJ}{mole}$.

Key equations for this chapter:

ENTROPY EQUATIONS:

$$\Delta S^o_{rxn} = \Sigma S^o_{f\,products} - \Sigma S^o_{f\,reactants}$$

$$\Delta S_{surr} = -\frac{\Delta H_{sys}}{T}$$

$$\Delta S_{universe} = \Delta S_{system} + \Delta S_{surroundings}$$

FOR A PHASE CHANGE:

$$\Delta S_{phase\,change} = \frac{\Delta H_{fusion}}{T} \quad \text{or} \quad \Delta S_{phase\,change} = \frac{\Delta H_{vaporization}}{T}$$

Section 9.1.1 Entropy of a Reaction, $\Delta S^o_{rxn} = \Sigma\, S^o_f \text{products} - \Sigma\, S^o_f \text{reactants}$

You can determine the entropy and spontaneity of any reaction by subtracting the sum of the entropy of the reactants from the sum of the entropy of the products. These problems are exactly like the calculation of the enthalpy, H, of a reaction. All values of S^o_f are found in standard thermodynamic tables.

Sample Problem 9.1 Entropy of a reaction

What is ΔS^o for the reaction:

$$2\, H_2O\,(l) \rightarrow 2\, H_2\,(g) + O_2\,(g)$$

Given:

S^o_f of $H_2O\,(l)$ = 69.95 $\frac{J}{K\,mole}$; S^o_f of $H_2\,(g)$ = 130.6 $\frac{J}{K\,mole}$; S^o_f of $O_2\,(g)$ = 205.0 $\frac{J}{K\,mole}$

Step 1. Identify the type of thermodynamic problem and the known and unknown variables.

In this problem, the problem asks you to determine the entropy of the reaction utilizing the entropy of formation for each component, S^o_f. Thus, to calculate entropy, you need this equation:

$$\Delta S^o_{rxn} = \Sigma\, S^o_f \text{products} - \Sigma\, S^o_f \text{reactants}$$

The known variables are the entropies of formation for each component, and the overall entropy for the reaction is unknown.

Step 2. Fill in the known variables, and solve for ΔS^o_{rxn}.

$$\Delta S^o_{rxn} = \Sigma\, S^o_f \text{products} - \Sigma\, S^o_f \text{reactants}$$

$$\Delta S^o_{rxn} = (S^o_f\, O_2 + 2 \times S^o_f\, H_2) - (2 \times S^o_f\, H_2O)$$

$$\Delta S^o_{rxn} = (205.0\,\tfrac{J}{K\,mole} + 2 \times 130.6\,\tfrac{J}{K\,mole}) - (2 \times 69.95\,\tfrac{J}{K\,mole})$$

$$\Delta S^o_{rxn} = 466.2\,\tfrac{J}{K\,mole} - 139.9\,\tfrac{J}{K\,mole} = +326.3\,\tfrac{J}{K\,mole}$$

Section 9.1.2 Predicting Spontaneity with $\Delta S_{universe} = \Delta S_{system} + \Delta S_{surroundings}$

In these problems, the entropy of the reaction is determined utilizing the ΔS_{system} and $\Delta S_{surroundings}$. The ΔS_{system} is calculated from this equation:

$$\Delta S^o_{rxn} = \Sigma S^o_{f\,products} - \Sigma S^o_{f\,reactants}$$

And the $\Delta S_{surroundings}$ from this equation: $\Delta S_{surr} = -\dfrac{\Delta H_{sys}}{T}$

Sample Problem 9.2 Predicting spontaneity of a reaction with entropy

Predict the spontaneity of the following reaction at 25°C:

$$2SO_2(g) + O_2(g) \rightarrow 2SO_3(g)$$

Given:

$\Delta H^o_{rxn} = -198$ kJ/mol

and

S^o_f of SO_3 (g) = 256.7 $\dfrac{J}{K\,mole}$; S^o_f of SO_2 (g) = 248.1 $\dfrac{J}{K\,mole}$; S^o_f of O_2 (g) = 205.0 $\dfrac{J}{K\,mole}$

Step 1. Identify the type of thermodynamic problem and the known and unknown variables.

When the enthalpy, H, of the reaction and the entropy of formation, S^o_f, are given, the equation needed to solve the problem is:

$$\Delta S_{universe} = \Delta S_{system} + \Delta S_{surroundings}$$

Where:

$$\Delta S^o_{system} = \Sigma S^o_{f\,products} - \Sigma S^o_{f\,reactants}$$

And

$$\Delta S_{surroundings} = -\dfrac{\Delta H_{sys}}{T}$$

Step 2. Calculate ΔS^o_{system} and $\Delta S_{surroundings}$.

Solving for ΔS^o_{system}:

$$\Delta S^o_{system} = \Sigma S^o_{f\,products} - \Sigma S^o_{f\,reactants}$$

$$\Delta S°_{system} = (2 \times SO_3(g)) - (2 \times SO_2(g) + O_2(g))$$

$$\Delta S°_{system} = (2 \times 256.7 \tfrac{J}{K\,mole}) - (2 \times 248.1 \tfrac{J}{K\,mole} + 205.0 \tfrac{J}{K\,mole})$$

$$\Delta S°_{system} = (513.4 \tfrac{J}{K\,mole}) - (701.2 \tfrac{J}{K\,mole})$$

$$\Delta S°_{system} = -187.8 \tfrac{J}{K\,mole}$$

Solving for $\Delta S_{surroundings}$

$$\Delta S_{surroundings} = -\tfrac{\Delta H_{sys}}{T}$$

The temperature is 25°C, which is 298.15 K.

$$\Delta S_{surroundings} = -\left(\tfrac{-198\ kJ/mole}{298.15\ K}\right)$$

$$\Delta S_{surroundings} = 0.66409 \tfrac{kJ}{mole\,K}$$

Converting kJ to J

$$\Delta S_{surroundings} = 664.1 \tfrac{J}{mole\,K}$$

Step 3. Calculate $\Delta S_{universe}$, and predict spontaneity.

$$\Delta S_{universe} = \Delta S_{system} + \Delta S_{surroundings}$$

$$\Delta S_{universe} = -187.8 \tfrac{J}{K\,mole} + 664.1 \tfrac{J}{mole\,K}$$

$$\Delta S_{universe} = 476.3 \tfrac{J}{mole\,K}$$

Spontaneous

Section 9.1.3 Predicting Entropy for a Phase Change

In these questions, the problem asks you to determine the entropy of a phase change in which you can use either one of these equations:

$$\Delta S_{phase\ change} = \tfrac{\Delta H_{fusion}}{T} \quad \text{or} \quad \Delta S_{phase\ change} = \tfrac{\Delta H_{vaporization}}{T}$$

Sample Problem 9.3 Predicting entropy for a phase change

What is the entropy change of 1 mole of liquid water turning to steam? The heat of vaporization for water is 40.65 kJ/mole.

Step 1. Identify the type of thermodynamic problem and the known and unknown variable.

In this problem, the entropy is known and the heat of vaporization is known. This thermodynamic problem, thus, requires this equation to calculate ΔS for a phase change:

$$\Delta S_{phase\ change} = \frac{\Delta H_{vaporization}}{T}$$

$$\Delta S_{vap} = \frac{\Delta H_{vaporization}}{T}$$

Step 2. Utilizing this equation, fill in the known variables, and solve for ΔS_{vap}.

$$\Delta S_{vap} = \frac{\Delta H_{vaporization}}{T}$$

The temperature at which water boils is 100°C, which is 373.15 K.

$$\Delta S_{vap} = \frac{40.65 \frac{kJ}{mole}}{373.15\ K}$$

$$\Delta S_{vap} = 0.1089\ \frac{kJ}{K\ mole}$$

Converting to the correct units for entropy (kJ to J), multiply by 1000).

$$\Delta S_{vap} = 108.9\ \frac{J}{K\ mole}$$

STUDENT PROBLEMS

1. The molar heat of fusion for aluminum is 8.66 $\frac{kJ}{mole}$. What is the entropy change? Aluminum melts at 660.3°C
2. Calculate ΔS^o_{rxn} for the reaction:

$$2\ C_4H_{10}\ (g) + 13\ O_2\ (g) \rightarrow 8\ CO_2\ (g) + 10\ H_2O\ (g)$$

S^o_f of $C_4H_{10}(g)$ = 310.1 $\frac{J}{K\ mole}$; S^o_f of CO_2 (g) = 213.7 $\frac{J}{K\ mole}$;

S^o_f of O_2 (g) = 205.0 $\frac{J}{K\ mole}$ and S^o_f of H_2O (g) = 188.8 $\frac{J}{K\ mole}$

3. Is this reaction spontaneous?

$$4\ Fe\ (s) + 3\ O_2\ (g) \rightarrow 2\ Fe_2O_3\ (s)$$

Given:

ΔH^o_f of Fe_2O_3 (s) = −825.5 $\frac{kJ}{mole}$; ΔH^o_f of O_2 (g) = 0.0 $\frac{kJ}{mole}$; ΔH^o_f of Fe(s) = 0.0 $\frac{kJ}{mole}$

and

S^o_f of Fe (s) = 27.3 $\frac{J}{K\ mole}$; S^o_f of O_2 (g) = 205.0 $\frac{J}{K\ mole}$; S^o_f of Fe_2O_3 (s) = 87.4 $\frac{J}{K\ mole}$ (Answers: 1. 9.28 J/K mole; 2. 312.4 J/K mole; 3. Yes)

Section 9.2 Gibbs Free Energy, G

Gibbs free energy is often considered a better predictor of the spontaneity. Different from entropy, when ΔG is negative, the reaction is spontaneous; when ΔG is positive, the reaction is non-spontaneous, and when ΔG = 0, the reaction is at equilibrium.

The set of equations used to solve Gibbs free energy problems are:

$$\Delta G^o_{rxn} = \Sigma\, G^o_{f\,products} - \Sigma\, G^o_{f\,reactants}$$

$$\Delta G^o = \Delta H^o - T\Delta S^o \text{ (standard)}$$

$$\Delta G = \Delta H - T\Delta S \text{ (nonstandard)}$$

$$\Delta G = \Delta G^o + RT \ln Q \text{ (nonstandard state)}$$

$$\Delta G^o = -RT \ln K \text{ (Standard State – Equilibrium)}$$

$$R = 8.314\, \frac{J}{K\,mole}; \quad T = \text{Kelvin}$$

Section 9.2.1 Gibbs Free Energy of a Reaction, $\Delta G^o_{rxn} = \Sigma\, G^o_{f\,products} - \Sigma\, G^o_{f\,reactants}$

You can determine the Gibbs free energy of any reaction by subtracting the sum of the Gibbs free energy of the reactants from the products. These problems are exactly like the calculation of the enthalpy, H, and entropy, S of a reaction. All values of G^o_f are found in standard thermodynamic tables.

Sample Problem 9.4 Gibbs Free Energy of a Reaction, $\Delta G^o_{rxn} = \Sigma\, G^o_{f\,products} - \Sigma\, G^o_{f\,reactants}$

<u>What is ΔG° for the reaction:</u>

$$2\, H_2O\,(l) \rightarrow 2\, H_2\,(g) + O_2\,(g)$$

Given:

G^o_f of H_2O (l) = -237.1 $\frac{kJ}{mole}$; G^o_f of H_2 (g) = 0.0 $\frac{kJ}{mole}$; G^o_f of O_2 (g) = 0.0 $\frac{kJ}{mole}$

Step 1. Identify the type of thermodynamic problem and the known and unknown variables.

In this problem, the Gibbs free energy of the reaction is asked to be determined utilizing the entropy of formation for each component, G_f^o. Thus, to calculate Gibbs free energy, you need the following equation:

$$\Delta G_{rxn}^o = \Sigma\, G_{f\,products}^o - \Sigma\, G_{f\,reactants}^o$$

The known variables are the Gibbs free energy of formation for each component, and the overall Gibbs free energy for the reaction is unknown.

Step 2. Fill in the known variables, and solve for ΔG_{rxn}^o.

$$\Delta G_{rxn}^o = \Sigma\, G_{f\,products}^o - \Sigma\, G_{f\,reactants}^o$$

$$\Delta G_{rxn}^o = (G_f^o\, O_2 + 2 \times G_f^o\, H_2) - (2 \times G_f^o\, H_2O)$$

$$\Delta G_{rxn}^o = (0.0\, \frac{kJ}{mole} + 2 \times 0.0\, \frac{kJ}{mole}) - (2 \times -237.1\, \frac{kJ}{mole})$$

$$\Delta G_{rxn}^o = 474.2\, \frac{kJ}{mole}$$

Section 9.2.2 Gibbs Free Energy of a Reaction; $\Delta G^o = \Delta H^o - T\Delta S^o$

The equation used to predict spontaneity of a reaction for Gibbs free energy is:

$$\Delta G^o = \Delta H^o - T\Delta S^o$$

Recall that when the value of ΔG^o is negative, the reaction is spontaneous; when ΔG^o is positive, the reaction is non-spontaneous, and when ΔG^o is zero, the reaction is in equilibrium. In many textbooks, the equilibrium condition will exist if the calculated value of ΔG^o is between $-10\, \frac{kJ}{mole}$ and $+10\, \frac{kJ}{mole}$.

Section 9.2.2.1 Type 1 Problem: Calculation of ΔG^o

Sample Problem 9.5 Calculation of ΔG^o from entropy and enthalpy

Calculate the value of Gibbs free energy and determine spontaneity for this reaction at 25°C.

$$2NO(g) + Cl_2(g) \rightleftharpoons 2NOCl(g)$$

Given:

$$\Delta H^o_{rxn} = -77.16 \frac{kJ}{mole}; \Delta S^o_{rxn} = -121.1 \frac{J}{K\,mole}$$

Step 1. Identify the type of thermodynamic problem and the known and unknown variables.

The problem asks to calculate Gibbs free energy, G, and provides the values for enthalpy and entropy. When all three thermodynamic values (G, H, S) are mentioned in a problem, this equation is used:

$$\Delta G^o = \Delta H^o - T\Delta S^o$$

In these problems, the temperature of the reaction is a necessary variable and needs to be in Kelvin.

Step 2. Fill in the known variables, and solve for ΔG^o.

Note: Before solving, make sure to change all units to either joules or kilojoules.

$$\Delta G^o = \Delta H^o - T\Delta S^o$$

$$\Delta H^o_{rxn} = -77.16 \frac{kJ}{mole}$$

$$\Delta S^o_{rxn} = -121.1 \frac{J}{mole} = -.1211 \frac{kJ}{K\,mole}$$

$$T = 25°C = 298.15\ K$$

Placing these values into the equation and solving for ΔG^o yields:

$$\Delta G^o = \Delta H^o - T\Delta S^o$$

$$\Delta G^o = -77.16 \frac{kJ}{mole} - (298.15\ K)(-.1211 \frac{kJ}{K\,mole})$$

$$\Delta G^o = -77.16 \frac{kJ}{mole} + 36.11 \frac{kJ}{mole}$$

$$\Delta G^o = -41.05 \frac{kJ}{mole}$$

SPONTANEOUS

Solving tip

In these types of problems, the question may also give the enthalpy and entropy of formation for each component. In these cases, the ΔH and ΔS of the reaction is calculated using:

$$\Delta S^o_{rxn} = \Sigma\, S^o_{f\,products} - \Sigma\, S^o_{f\,reactants}$$

or

$$\Delta H^o_{rxn} = \Sigma\, H^o_{f\,products} - \Sigma\, H^o_{f\,reactants}$$

Section 9.2.2.2 Type 2 Problem: Calculation of ΔG° at Different Temperatures

Sample Problem 9.6 Calculation of ΔG° at different temperatures

There is a temperature dependence on Gibbs free energy, which means the value will change with varying temperatures. However, enthalpy and entropy do not have this temperature dependence, and their values do not change. Thus, the calculated ΔH and ΔS from thermodynamic tables holds for any temperature cited in these problems.

Sample Problem

For this reaction at 298 K:

$$N_2\,(g) + 3\,H_2\,(g) \rightleftharpoons 2\,NH_3(g)$$

$\Delta S° = -198\,\frac{J}{K\,mole}$, and $\Delta H° = -91.8\,\frac{kJ}{mole}$. What is the value of ΔG° at 850 K?

Step 1. Identify the type of thermodynamic problem and the known and unknown variables.

The problem asks to calculate Gibbs free energy, G, and provides the values for enthalpy and entropy. When all three thermodynamic values (G, H, S) are mentioned in a problem, this equation is used:

$$\Delta G° = \Delta H° - T\Delta S°$$

The only difference from the prior problem is that the question asks for ΔG° at 850 K, whereas the data given is at 298K.

Step 2. Solve for ΔG° at 850K.

In order to solve this problem, you must enter the values for ΔH°, ΔS°, and the 850K temperature into:

$$\Delta G° = \Delta H° - T\Delta S°$$

Make sure to match units! Convert J to kJ for entropy.

$$\Delta G° = -91.8 \frac{kJ}{mole} - (850 \text{ K})(-.198 \frac{kJ}{K \text{ mole}})$$

$$\Delta G° = -91.8 \frac{kJ}{mole} + 168.3 \text{ K} \frac{kJ}{K \text{ mole}}$$

$$\Delta G° = 76.5 \frac{kJ}{mole}$$

Section 9.2.2.3 Type 3 Problem: Calculation of Temperature for Reaction Spontaneity

Sample Problem 9.7 Calculation of temperature for reaction spontaneity

In these problems, the question asks, "at what temperature does the reaction become spontaneous?" This means that, at this temperature, an equilibrium exists, and $\Delta G° = 0$, and the equation:

$$\Delta G° = \Delta H° - T\Delta S°$$

Goes to

$$0 = \Delta H° - T\Delta S°$$

With algebraic manipulation, this transition temperature is expressed as:

$$T = \frac{\Delta H°}{\Delta S°}$$

<u>At what temperature</u> does the reaction become spontaneous:

$$N_2 (g) + 3 H_2 (g) \rightleftarrows 2 NH_3(g)$$

$$\Delta S° = -198 \frac{J}{K \text{ mole}} \text{ and } \Delta H° = -91.8 \frac{kJ}{mole} \text{ and } \Delta G° = -32.8 \frac{kJ}{mole}$$

Step 1. Identify the type of thermodynamic problem and the known and unknown variables.

The question asks for the temperature at which the reaction transitions, from non-spontaneous to spontaneous. The temperature is unknown, but the ΔH and ΔS are known. $\Delta G°$ is given but is not necessary to solve this problem, because:

$$T = \frac{\Delta H°}{\Delta S°}$$

Step 2. Fill in known variables, and solve for temperature.

To start, make sure the units match; thus, change S from J to kJ.

$$T = \frac{-91.8 \frac{kJ}{mole}}{-0.198 \frac{kJ}{K \, mole}}$$

$$T = 463.6 \text{ K}$$

Section 9.2.3 Gibbs Free Energy and Equilibrium Constants

The set of equations used to solve these types of thermodynamic problems are:

$$\Delta G = \Delta G° + RT \ln Q \text{ (nonstandard state)}$$

$$\Delta G° = - RT \ln K \text{ (Standard State – Equilibrium)}$$

$$R = 8.314 \frac{J}{K \, mole}; \quad T = \text{Kelvin}$$

These problems ask to predict spontaneity or calculate $\Delta G°$ given K or calculate K given $\Delta G°$.

$\Delta G°$ can be given or may have to be calculated from given $\Delta H°$ and $\Delta S°$ values using:

$$\Delta G° = \Delta H° - T\Delta S°$$

Sample Problem 9.8 Gibbs free energy and equilibrium constants

The K_c at 450°C for the following reaction is 0.020. What is $\Delta G°$?

$$2HI(g) \rightleftarrows H_2(g) + I_2(g)$$

Step 1. Identify the type of thermodynamic problem and the known and unknown variables.

This problem gives K_c and asks to calculate $\Delta G°$; thus, you must use the following equation:

$$\Delta G° = - RT \ln K$$

Step 2. Fill in the known variables, and solve for $\Delta G°$.

$$\Delta G° = - RT \ln K$$

$$\Delta G° = - (8.314 \frac{J}{K \, mole})(723.15 \text{ K}) \ln (0.02)$$

$$\Delta G° = 2.35 \times 10^4 \frac{J}{mole} = 23.5 \frac{kJ}{mole}$$

Note: Make sure to convert temperature to Kelvin, and convert the final answer for $\Delta G°$ from J to kJ.

Sample Problem 9.9 Gibbs free energy and equilibrium constants

<u>Calculate ΔG for the reaction at 25°C:</u>

$$N_2 (g) + 3 H_2 (g) \rightleftharpoons 2 NH_3(g)$$

Given

$$\Delta S° = -198 \frac{J}{K \, mole} \text{ and } \Delta H° = -91.8 \frac{kJ}{mole} \text{ and } \Delta G° = -32.8 \frac{kJ}{mole}$$

And the partial pressures of P_{NH_3} = 2.0 atm; P_{H_2} = 4.6 atm; and P_{N_2} = 1.4 atm

Step 1. Identify the type of thermodynamic problem and the known and unknown variables.

In this question, the problem asks for the non-standard value of ΔG. Thus, the equation necessary to solve this problem is:

$$\Delta G = \Delta G° + RT \ln Q$$

You have the $\Delta G°$ and the temperature, as well as partial pressures of each component. These partial pressures will be used to calculate the reaction quotient, Q.

Step 2. Determine Q.

$$Q = \frac{[NH_3]^2}{[N_2][H_2]^3}$$

Fill in the partial pressure:

$$Q = \frac{[2.0 \, atm]^2}{[1.4 \, atm][4.6 \, atm]^3}$$

$$Q = 0.02935$$

Step 3. Fill in the known variables, and solve for ΔG, and predict spontaneity.

$$\Delta G = \Delta G° + RT \ln Q$$

Make sure to convert $\Delta G°$ from kJ to J to match units in the gas constant, R.

$$\Delta G = -32{,}800 \frac{J}{mole} + (8.314 \frac{J}{K \, mole})(298.15 \, K) \ln (0.02935)$$

$$\Delta G = -32{,}800 \frac{J}{mole} + (8.314 \frac{J}{K \, mole})(298.15 \, K) \ln (0.02935)$$

$$\Delta G = -32{,}800 \ \frac{J}{mole} - 8746.4 \ \frac{J}{mole}$$

$$\Delta G = -41546.4 \ \frac{J}{mole} = -41.6 \ \frac{kJ}{mole}$$

STUDENT PROBLEMS

1. What is $\Delta G°$ for the reaction $2NO(g) + Cl_2(g) \rightleftarrows 2NOCl(g)$ if K_p is 6.5×10^{-4} at 300 K?
2. What is K_c at 500°C for the reaction $3H_2(g) + N_2(g) \rightleftarrows 2NH_3(g)$ if $\Delta G° = -25.5 \ \frac{kJ}{mole}$?
3. What is ΔG for the reaction $CO_2(g) + C(s) \rightleftarrows 2 CO(g)$ at 1000°C if the concentration of $[CO_2] = 0.03 M$ and $[CO] = 0.11 M$? If at 25°C, $\Delta H° = 172.5 \ \frac{kJ}{mole}$, $\Delta S° = 181.3 \ \frac{J}{mole}$.
4. For the reaction $WO_3(s) + 3H_2(g) \rightleftarrows W(s) + 3H_2O(g)$ at 25°C, $\Delta G° = 77.3 \ \frac{kJ}{mole}$; $\Delta H° = 114.5 \ \frac{kJ}{mole}$. What is the $\Delta S°$ for the reaction? At what temperature does the reaction become spontaneous?
5. What is the $\Delta G°$ at 2000 K for the reaction:

$$C_3H_8(g) + 5O_2(g) \rightleftarrows 3 CO_2(g) + 4 H_2O(g)$$

If at 25°C; $\Delta H° = -2215 \ \frac{kJ}{mole}$, $\Delta S° = 101 \ \frac{J}{mole}$

(Answers: 1. -18.3 kJ/mol; 2. 52.9; 3. -9,670 kJ/mole; 4. $\Delta S° = 125$ J/K mole, T = 917 K; 5. -2,417 kJ/mole)

Chapter 10. Electrochemistry

Relevant Equations for Electrochemistry

$$E^o_{cell} = E^o_{cathode} - E^o_{anode}$$

$$\Delta G^o = -nF E^o_{cell}$$

$$\Delta G^o = -RT \ln K$$

$$E^o_{cell} = \frac{RT}{nF} \ln K$$

simplified at 25°C (R = 8.314 J/K mol, F = 96,485 C/mole e⁻) becomes

$$E^o_{cell} = \frac{0.0592\ V}{n} \log K$$

NERNST EQUATION:

$$E_{cell} = E^o_{cell} - \frac{RT}{nF} \ln Q$$

simplified at 25°C (R = 8.314 J/K mol, F = 96,485 C/mole e⁻) becomes

$$E_{cell} = E^o_{cell} - \frac{0.0592\ V}{n} \log Q$$

ELECTROLYSIS:

Grams Electrolyzed $= \dfrac{\text{Current x time}}{nF}$ **x Molar Mass**

Current = Ampere/sec; time = sec, n = moles of electrons; F = 96,485 C/mole e⁻

Volt, V = Joule/Coulomb

Cathode = Reduction Occurs

Anode = Oxidation Occurs

Section 10.1 Galvanic Cells: Cell Potential, E^o_{cell} or Electromotive Force, EMF

The standard cell potential, EMF, or electrode potential, is when the solute concentrations are 1 M, the gas pressure is 1 atm, and the temperature is 25°C. To calculate the cell potential, the standard electrode potential of the anode is subtracted from the cathode. The equation that depicts this is simply:

$$E^o_{cell} = E^o_{cathode} - E^o_{anode}$$

You have three ways to determine the cell potential in these problems; you can determine it by if the question defines which is the cathode and the anode via a chemical equation; you can determine it by electrochemical cell notation; or the student must determine the cathode and anode reaction by looking at the Standard Electrode Potential Chart. A short list is shown below; however, a complete table is typically found within the Electrochemistry Chapter or as an appendix of any general chemistry book.

Table 10.1 Standard Potentials

Cathode (Reduction) Half Reaction	Standard Potential E^o (V)
Li^{+1} (aq) + e^- → Li (s)	-3.04
Al^{+3} (aq) + 3 e^- → Al (s)	-1.66
Zn^{+2} (aq) + 2 e^- → Zn (s)	-0.762
Pb^{+2} (aq) + 2 e^- → Pb (s)	-0.126
Cu^{+1} (aq) + e^- → Cu (s)	0.521
Ag^{+1} (aq) + e^- → Ag (s)	0.799

Sample Problem 10.1 Calculation of overall cell potential – standard potential values

What is the overall cell potential for:

$$Fe\ (s) + Cu^{+2}\ (aq) \rightarrow Fe^{+2}\ (aq) + Cu\ (s)$$

E^o_{Fe} = -0.41 V; E^o_{Cu} = 0.34 V

Step 1. Identify the type of cell potential problem and the known and unknown quantities.

The overall cell potential is:

$$E°_{cell} = E°_{cathode} - E°_{anode}$$

Therefore, you must identify the cathode and anode. The known variables are the standard cell potentials for iron and copper.

Step 2. Identify the cathode and anode; solve for $E°_{cell}$.

Since a chemical reaction is given; it dictates which is the anode and which is the cathode. In all cases, you need to identify only one of the half-reactions, because the other is the opposite reaction. For this problem:

$$Fe \rightarrow Fe^{+2}$$

Adding the electrons to balance the reaction yields

$$Fe \rightarrow Fe^{+2} + 2\ e^-$$

This reaction loses electrons and is the OXIDATION reaction. This occurs at the ANODE.

Thus, the Cu half reaction occurs at the CATHODE.

Solving for $E°_{cell}$:

$$E°_{cell} = E°_{cathode} - E°_{anode}$$

$$E°_{cell} = E°_{Cu} - E°_{Fe}$$

$$E°_{cell} = 0.34\ V - (-0.41V)$$

$$E°_{cell} = 0.75\ V$$

Sample Problem 10.2 Calculation of overall cell potential – cell notation

What is the overall cell potential for:

$$Zn\ (s)|Zn^{+2}\ (aq)||Cr^{+3}|Cr(s)?$$

$E°_{Zn} = -0.76\ V;\ E°_{Cr} = -0.74\ V$

Step 1. Identify the type of cell potential problem and the known and unknown quantities.

The overall cell potential is:

$$E°_{cell} = E°_{cathode} - E°_{anode}$$

Thus, the cathode and anode must be identified. The known variables are the standard cell potentials for zinc and chromium.

Step 2. Identify the cathode and anode; solve for E^o_{cell}.

Since the standard cell notation is provided, the Zn is the anode half-reaction, while the Cr half reaction occurs at the cathode.

Solving for E^o_{cell}:

$$E^o_{cell} = E^o_{cathode} - E^o_{anode}$$

$$E^o_{cell} = E^o_{Cr} - E^o_{Zn}$$

$$E^o_{cell} = -0.74\ V - (-0.76\ V)$$

$$E^o_{cell} = 0.02\ V$$

Sample Problem 10.3 Calculation of overall cell potential – no reaction, standard potential

What would the overall cell potential be if Pb^{+2}/Pb 1M was combined in a galvanic cell with Mg^{+2}/Mg 1M? $E^o_{Pb} = -0.13\ V$; $E^o_{Mg} = -2.38\ V$.

Step 1. Identify the type of cell potential problem and the known and unknown quantities.

The overall cell potential is:

$$E^o_{cell} = E^o_{cathode} - E^o_{anode}$$

Thus, the cathode and anode must be identified. The known variables are the standard cell potentials for lead and magnesium.

Step 2. Identify the cathode and anode; solve for E^o_{cell}.

In this problem, the chemical reaction nor the standard cell notation are given. Thus, the anodic and cathodic reaction must be determined simply by the standard half-cell potentials provided in the problem.

In all cases, the more **POSITIVE** potential associated with the half-reaction will be the reduction reaction and occur at the **CATHODE**. For this problem:

$$E^o_{Pb} = -0.13\ V;\ E^o_{Mg} = -2.38\ V$$

And, the lead possesses the more positive value -0.13 and, thus, is the CATHODE.

Solving for E^o_{cell}:

$$E^o_{cell} = E^o_{cathode} - E^o_{anode}$$

$$E^o_{cell} = E^o_{Pb} - E^o_{Mg}$$

$$E^o_{cell} = -0.13 \text{ V} - (-2.38 \text{V})$$

$$E^o_{cell} = 2.25 \text{ V}$$

STUDENT PROBLEMS

1. What is the overall cell potential for Zn^{+2}/Zn combined with Ca^{+2}/Ca?
2. What is the overall cell potential for the galvanic cell with this reaction?

$$Au \text{ (s)} + Al^{+3} \text{ (aq)} \rightarrow Au^{+3} \text{ (aq)} + Al \text{ (s)}$$

3. What is the overall cell potential for the galvanic cell: K, K^{+1} || Ni^{+2}, Ni?

(Answers: 1. 2.00 V; 2. -3.16 V; 3. 2.69 V)

Section 10.2 Gibbs Free Energy and Cell Potential

By this set of equations:

$$\Delta G^o = -nF\, E^o_{cell}$$

$$\Delta G^o = -RT \ln K$$

$$E^o_{cell} = \frac{RT}{nF} \ln K$$

simplified at 25°C (R = 8.314 J/K mol, F = 96,485 C/mole e⁻) becomes

$$E^o_{cell} = \frac{0.0592\, V}{n} \log K$$

There is a direct relationship between Gibbs free energy, ΔG^o; the equilibrium constant, K, and the overall cell potential, E^o_{cell}.

Thus, the spontaneity of a cell can be predicted by looking at the sign of either ΔG^o or E^o_{cell}.

This table will assist in the determination of spontaneity.

Table 10.2 Gibbs Free Energy and Spontaneity Rules

	Spontaneous	Non-Spontaneous
ΔG^o	Negative	Positive
E^o_{cell}	Positive	Negative
K	>1	<1

Sample Problem 10.4 Gibbs free energy and cell potential

Calculate the value of E^o_{cell}, ΔG^o, and K for this reaction at 25°C:

$$2Au^{+3}(aq) + 3Ca(s) \rightarrow 2Au(s) + 3Ca^{+2}(aq)$$

With Au^{+3}/Au, E^o = 1.50 V and Ca^{+2}/Ca, E^o = -2.76 V

Step 1. Identify the type of cell potential problem and the known and unknown quantities.

There are three parts you must solve for in this problem: E^o_{cell}, ΔG^o, and K.

Since the equations for $\Delta G°$ and K need $E°_{cell}$, you must determine it first by the equation:

$$E°_{cell} = E°_{cathode} - E°_{anode}$$

Thus, you must identify the cathode and anode. The known variables are the standard cell potentials for gold and calcium.

Next, you can solve for $\Delta G°$ and K utilizing:

$$\Delta G° = -nF\, E°_{cell}$$

$$\Delta G° = -RT \ln K$$

Step 2. Identify the cathode and anode, and solve for $E°_{cell}$.

Since a chemical reaction is given; it dictates what the anode is and which is the cathode. In all cases, you only need to identify one of the half-reactions, because the other is the opposite reaction. For this problem:

$$Au^{+3} \rightarrow Au$$

Adding the electrons to balance the reaction yields

$$Au^{+3} + 3\,e^- \rightarrow Au$$

This reaction gains electrons, is the REDUCTION reaction, and will occur at the CATHODE.

Thus, the Ca half reaction occurs at the ANODE.

Solving for $E°_{cell}$:

$$E°_{cell} = E°_{cathode} - E°_{anode}$$

$$E°_{cell} = E°_{Au} - E°_{Ca}$$

$$E°_{cell} = 1.50\text{ V} - (-2.76\text{ V})$$

$$E°_{cell} = 4.26\text{ V}$$

Step 3. Determine n, the number of moles of electrons, and solve for ΔG.

Next $\Delta G°$ is calculated with:

$$\Delta G° = -nF\, E°_{cell}$$

For this equation, F = 96,485 C/mole e⁻, E^o_{cell} is 4.26V, and n, the number of moles of electrons, needs to be calculated before we can solve for ΔG°. To determine n, you need the balance equation:

$$2Au^{+3}(aq) + 3Ca(s) \rightarrow 2Au(s) + 3Ca^{+2}(aq)$$

Typically, these REDOX reactions have been balanced, and to determine n, just look at either one of the ions in this equation. In this case, multiply the coefficient prior to the ion times its charge, thus giving the value of n. For this problem:

$$3Ca^{+2} = 3 \times 2 = 6$$

$$n = 6$$

Solving for ΔG

$$\Delta G^o = -nF\,E^o_{cell}$$

$$\Delta G^o = -(6 \text{ mole } e^-)(96,485 \text{ C/mole } e^-)(4.26 \text{ V})$$

$$\Delta G^o = -2.47 \times 10^6 \text{ Joules} = -2.47 \times 10^3 \text{ kJ}$$

(recall: Coulomb × volt = Joules)

Step 4. Solve for the equilibrium constant, K.

Since ΔG is now known, K is solved for by:

$$\Delta G^o = -RT \ln K$$

Gibbs free energy must be in joules to match the joules unit in the gas constant, R.

$$-2.47 \times 10^6 \text{ Joules} = -(8.314 \text{ Joules/mole K})(298.15 \text{ K}) \ln K$$

$$\ln K = \frac{2.47 \times 10^6 \text{ Joules}}{(8.314 \text{ Joules/mole K})(298.15 \text{ K})}$$

$$\ln K = \frac{2.47 \times 10^6 \text{ Joules}}{(8.314 \text{ Joules/mole K})(298.15 \text{ K})}$$

$$\ln K = 996.44$$

$$K = e^{996.44}$$

$$K = 0.0 \text{ (extremely small K)}$$

In this sample problem, all three variables were calculated. Typically, these types of problems can ask for either, both, or all three quantities. You will always utilize the same set of equations to solve these problems.

STUDENT PROBLEMS

1. If $K = 1.2 \times 10^5$, calculate E°_{cell} and ΔG° for the reaction at 25°C of Ni/Ni^{+2} with Cd/Cd^{+2} with Ni/Ni^{+2}.

2. Calculate the value of E°_{cell}, ΔG°, and K for this reaction at 25°C:

$$2Fe^{+3} (aq) + 3Sn (s) \rightarrow 2Fe (s) + 3Sn^{+2} (aq)$$

With Fe^{+3}/Fe, $E^\circ = -0.04$ V and Sn/Sn^{+2}, $E^\circ = -0.14$ V

(Answers: 1. $\Delta G^\circ = -29.0$ kJ/mole, $E^\circ_{cell} = 0.15$ V; 2. $E^\circ_{cell} = 0.1$ V, $\Delta G^\circ = -57.9$ kJ/mol, $K = 1.40 \times 10^{10}$)

Section 10.3 Nernst Equation, Non-Standard Conditions

Utilize the Nernst equation to solve problems for which there are non-standard state conditions in galvanic cells (solutions are not 1 M). In these cases, the sign of E_{cell} is used to see if the galvanic cell will be spontaneous (positive E) and produce a charge or non-spontaneous (negative E). The equation utilized in these questions is:

$$E_{cell} = E°_{cell} - \frac{RT}{nF} \ln Q$$

or simplified at 25°C (R = 8.314 J/K mol, F = 96,485 C/mole e⁻) becomes

$$E_{cell} = E°_{cell} - \frac{0.0592 \text{ V}}{n} \log Q$$

These problems are readily identifiable, because concentrations are either given or unknown quantities for the components at the anode or cathode.

Sample Problem 10.5 Nernst equation, non-standard conditions

If the $E°_{cell}$ of the following reaction is 0.61 V. Will the reaction be spontaneous at 25°C if the concentration of [Pb^{+2}] = 0.01 M and [Cr^{+3}] = 0.35 M?

$$2Cr(s) + 3Pb^{2+}(aq) \rightarrow 3Pb(s) + 2Cr^{3+}(aq)$$

Step 1. Identify the type of cell potential problem and the known and unknown quantities.

Since concentrations are given, this problem requires solution with the Nernst equation. Since the temperature is 25°C, you can use this form of the equation:

$$E_{cell} = E°_{cell} - \frac{0.0592 \text{ V}}{n} \log Q$$

$E°_{cell}$, the concentrations of the anode and cathode are given. You do not know the value of E, Q, and n.

Step 2. Determine the value of n, moles of electrons, and the reaction quotient, Q.

To determine n, you need the balance equation:

$$2Cr(s) + 3Pb^{+2}(aq) \rightarrow 3Pb(s) + 2Cr^{+3}(aq)$$

Typically, these REDOX reactions have been balanced, and to determine n, just look at either one of the ions in this equation. In this case, multiply the coefficient of the ion in the balanced equation times the charge on the ion which yields the value of n. For this problem:

$$3Pb^{+2} = 3 \times 2 = 6$$

$$n = 6$$

To determine Q, you must write the equilibrium expression for the reaction:

$$2Cr(s) + 3Pb^{+2}(aq) \rightarrow 3Pb(s) + 2Cr^{+3}(aq)$$

$$Q = \frac{[Cr^{+3}]^2}{[Pb^{+2}]^3}$$

(Recall: solids and liquids are excluded from the equilibrium expression)

$$Q = \frac{(0.35)^2}{(0.01)^3}$$

$$Q = 1.225 \times 10^5$$

Step 3. Solve for E_{cell}, and determine spontaneity.

$$E_{cell} = E^o_{cell} - \frac{0.0592\ V}{n} \log Q$$

$$E_{cell} = 0.61\ V - \frac{0.0592\ V}{6} \log (1.225 \times 10^5)$$

$$E_{cell} = 0.56\ V$$

E_{cell} is positive and the reaction is spontaneous

STUDENT PROBLEMS

1. The overall cell potential for the galvanic cell with this reaction:

$$Au\ (s) + Al^{+3}\ (aq) \rightarrow Au^{+3}\ (aq) + Al\ (s)$$

    is -3.16 V. Is this reaction spontaneous at 25°C, if the concentrations of $Al^{+3} = 1.1 \times 10^{-3}$ and $Au^{+3} = 0.50$ M?

2. Calculate the cell potential for the following galvanic cell at 25°C:
$$Na/Na^+(0.10\ M)||Ag^+/Ag\ (2.0M)$$

3. $E^o_{cell} = 0.75$ V at 25°C for $Fe\ (s) + Cu^{+2}\ (aq) \rightarrow Fe^{+2}\ (aq) + Cu\ (s)$. What is the concentration of Fe^{+2} if the $E_{cell} = 0.68$ V and $[Cu^{+2}] = 0.0015$ M?

(Answers: 1. E = -3.28 V non-spontaneous; 2. 3.43 V; 3. 0.016 M)

# Chapter 10. Electrochemistry

Section 10.4 Electrolysis Calculations

Electrolysis is a process by which an electric current passes through a solution, and the ions are deposited from the solution. In these problems, you can calculate the gram amount of the ion of interest utilizing the following equation:

$$\text{Grams Electrolyzed} = \frac{\text{Current} \times \text{time}}{n\,F} \times \text{Molar Mass}$$

Where the current = ampere (Coulomb/sec); time = sec, n = moles of electrons;

F = 96,485 C/mole e⁻.

Sample Problem 10.6 Electrolysis calculations

How many grams of gold (Au^{+3}) can be electrolyzed when a current of 8.8 A is passed through the gold solution for 90 minutes?

Step 1. Identify the type of cell potential problem and the known and unknown quantities.

In this problem, the question specifically mentions electrolysis. Thus, you should use the following equation:

$$\text{Grams Electrolyzed} = \frac{\text{Current} \times \text{time}}{n\,F} \times \text{Molar Mass}$$

You know the current, time, and metal; you do not know the quantity of gold electrolyzed.

Step 2. Solve for grams of gold.

Molar mass of Au = 196.97 grams/mole

90 minutes = 90 x 60 = 5400 sec

n = moles of electrons for Au^{+3} + 3 e⁻ → Au; thus n = 3

$$\text{Grams Electrolyzed} = \frac{\text{Current} \times \text{time}}{n\,F} \times \text{Molar Mass}$$

$$\text{Grams Electrolyzed} = \frac{(8.8\,\tfrac{C}{s})(5400\text{ sec})}{(3)(96{,}485\,\tfrac{C}{\text{mole e}^-})} \times (196.97\text{ grams/mole})$$

Grams Electrolyzed = 32.34 grams Au

STUDENT PROBLEMS

1. How many grams of copper are electroplated from a solution of $CuSO_4$, if a current of 12.5 A is passed through the solution for 20.0 hours?

2. How many hours are needed to generate 125 grams of Ni from a solution of NiSO₄ that undergoes a current of 5.00 A?

(Answers: 1. 296 grams of Cu; 2. 22.8 hours)

Printed by Libri Plureos GmbH in Hamburg, Germany